RESIDENTIAL WATER DEMAND
Alternative choices for management

UNIVERSITY OF TORONTO DEPARTMENT
OF GEOGRAPHY RESEARCH PUBLICATIONS

1. THE HYDROLOGIC CYCLE AND THE WISDOM OF GOD: A THEME IN
    GEOTELEOLOGY by Yi-Fu Tuan

2. RESIDENTIAL WATER DEMAND AND ECONOMIC DEVELOPMENT by
    Terence R. Lee

3. THE LOCATION OF SERVICE TOWNS: AN APPROACH TO THE
    ANALYSIS OF CENTRAL PLACE SYSTEMS by John U. Marshall

4. KANT'S CONCEPT OF GEOGRAPHY AND ITS RELATION TO RECENT
    GEOGRAPHICAL THOUGHT by J. A. May

5. THE SOVIET WOOD-PROCESSING INDUSTRY: A LINEAR PROGRAMMING
    ANALYSIS OF THE ROLE OF TRANSPORTATION COSTS IN LOCATION
    AND FLOW PATTERNS by Brenton M. Barr

6. THE HAZARDOUSNESS OF A PLACE: A REGIONAL ECOLOGY OF
    DAMAGING EVENTS by Kenneth Hewitt and Ian Burton

7. RESIDENTIAL WATER DEMAND: ALTERNATIVE CHOICES FOR
    MANAGEMENT by Angelo P. Grima

RESIDENTIAL WATER DEMAND
Alternative choices for management

Angelo P. Grima

Toronto & Buffalo
1972

Published for the University of Toronto
Department of Geography
by the University of Toronto Press

© University of Toronto Department of Geography
Published by University of Toronto Press
Toronto and Buffalo, 1972
Printed in Canada
ISBN 0-8020-3290-7
Microfiche ISBN 0-8020-0237-4

Acknowledgements

During the writing of this study I have been fortunate to receive the assistance and advice of many persons and organizations. The water revenue offices of several municipalities in the study area made it possible for me to obtain a sample of householders from their billing records for 1967. Water supply engineers and billing supervisors answered my questions about technical details of water supply problems and the pricing of water.

Special thanks are due to Professor Ian Burton who acted as my supervisor; he kept a close interest in every stage of the work and I am grateful for his critical appraisal and friendly encouragement.

Professor Y. Kotowitz (Department of Political Economy) and Professor J. N. H. Britton (Department of Geography) read successive versions of the draft; Professors F. Kenneth Hare, J. B. R. Whitney, J. Simmons, G. Jump, P. H. Jones, G. F. White, R. W. Kates, S. T. Wong, and Dr. T. R. Lee read the final draft of this study. Their helpful comments are appreciated. I have also benefited from the advice of other faculty members of the University of Toronto.

I would also like to thank Mr. S. Schulte who advised me on the use of the computer facilities in the University, and Miss Jennie Wilcox and Mr. G. Matthews who drew the figures, and Mrs. Susan Britton who made several helpful suggestions in the final editing of the manuscript.

The Canadian Commonwealth Scholarship and Fellowship Administration gave me financial support during my stay in Toronto.

Throughout the writing of this study, my wife Gemma continually supplied a very welcome understanding and encouragement.

Toronto,
1971   Angelo P. Grima

Many things change, many remain the same.

First Century:

Will anyone compare the idle Pyramids, or those other useless though much renowned works of the Greeks with these ... many indispensable structures ?

> Frontinus, De Aquis Urbis Romae, A.D. 97
> (translated by C. Herschel, 1899).

Twentieth Century:

If a single program were chosen which would have the maximum health benefits, which would rapidly stimulate social and economic development, and which would materially improve the standard of living of people, that program would be water supply with provision for water running into or adjacent to the house.

> Pan American Health Organization, Pan
> American Sanitary Bureau — Regional
> Office of the World Health Organization,
> Facts on Health Problems (Washington, DC:
> July 1961), p. 29.

Contents

ix

Tables

Illustrations

RESIDENTIAL WATER DEMAND
Alternative choices for management

I

Scope and organization
of the analysis

A BASIC MAN-NATURE TRANSACTION

The development of a water supply, one of the most basic trans-
actions between man and nature, has effects that are etched not
only in man-made landscape features such as aqueducts, reservoirs, dams, pipelines, treatment plants, and water towers but
also in water laws, customs, property rights, and folklore. The
complexity of this transaction depends on the characteristics of
the environment in which it occurs; of importance are the size
and distribution of the population, its rate of growth, level of
income, quality of housing, use of water-using appliances, the
state of technology, the administrative organization of the
community, and the availability, quality, and distribution of
surface and ground water resources.

In a culture where advanced technology is not available a
nearby source of water is a necessary condition for urban growth;
under such conditions "the presence of water... [is]... a con-
trolling factor in the distribution of population."[1] Technological

[1]Ellen C. Semple, "Domestic and Municipal Waterworks in Ancient Mediterranean Lands,"
Geographical Review 21, no. 4 (1931):466.

3

progress removes this constraint on urban growth provided the community is able and willing to devote scarce resources (land, labour, and materials) to the water supply projects. As the resources committed to the development of a community water supply could have been used to provide other goods and services the opportunity cost of extending the urban water supply system is likely to be high at a time of rapid metropolitan growth. There is pressure on public expenditures, not only for water supply[2] and sewage works, but also health, recreation, and other municipal services.

The magnitude of future municipal water needs is measured by the vast scale of urban population growth in the world as a whole: in 1962 there were about 1200 million people living in cities; the estimate for 1980 is 1700 million (an increase of 42 per cent) and for the year 2000 the estimate is 2500, or more than double the figure for 1962.[3] The provision of municipal water to meet the demand of the projected urban populations during the next few decades will require large-scale investment which is almost wholly irreversible. Furthermore, unless new technological breakthroughs are made and adopted, the level of investment is likely to increase faster than the growth in urban population, due to a higher per capita rate of use and the increasing cost of developing water resources as the more accessible sources prove to be insufficient. In general the emphasis in water utility management has been to develop new sources of municipal water in order to meet projected 're-quirements' for municipal water. This emphasis on the 'supply fix' disregards the efficiency principle that investment in water resources for whatever purpose should be increased up to the point where it is justified by the value that the community puts on it.

In the past the satisfaction contributed by a dependable supply of pure, piped water probably outweighed the cost. However,

[2]Following N. Wollman, Water Supply and Demand: Preliminary Estimates for 1980 and 2000, US Congress, Senate Select Committee on National Water Resources, 86th Cong., 2nd Sess., Comm. Print No. 32 (Washington DC: GPO, August 1960) p. 2 footnote 5, the terms 'supply' and 'demand' are used in the sense of physical requirements or amounts used (demand), and potential physical availability based on the capacity of the water treatment plant plus storage (supply). The text will make it clear when the same terms are used in the economists' sense of a schedule relating price and quantity.

[3]Bernd H. Dieterich and J. M. Henderson, Urban Water Supply Conditions and Needs in Seventy-Five Developing Countries, World Health Organization Public Health Papers No. 23 (Geneva: WHO, 1963), p. 13.

4

the policy of supplying <u>more</u> water at a higher cost needs to be appraised in view of the increasing per capita use particularly during periods of peak demand. The last few gallons demanded by the consumers are less 'essential' and yield less satisfaction than the first few gallons used for drinking, cooking, and washing. The additional expenditures required to develop water resources for municipal purposes have to be justified in terms of the value that consumers put on such water. This value-oriented policy is advocated in a recent report of the Committee on Water of the National Academy of Sciences — National Research Council. Enunciating some principles that "merit more attention than they are now receiving, " its Chairman notes:

In particular, the Committee, recognizing that the value of water varies among different groups in different places and at different times, believes that the management of water resources has evolved to a stage where planning should center upon the needs of the people rather than of water <u>per se</u>. This viewpoint implies that a broad range of alternatives must be considered before a decision is made to develop a water resource.[4]

In planning municipal water resources there is a set of such alternatives which may be conveniently grouped under the heading of demand engineering or demand management;[5] a subset of this group of alternatives is related to the price of residential water — i.e., the value that the consumers put upon water used for a variety of purposes in and around the house. An analysis of user behaviour with respect to the price of water he uses would indicate whether and how much further investment in this basic, capital-intensive municipal service is justified in terms of efficient allocation of the community's scarce resources.

The level of residential water use and the level of the investment requirements to develop new sources of supply can be reduced by distinguishing between the essential (or more highly valued), and the less essential uses of water in and around the house. Arbitrary restrictions can be applied on the use of water for certain purposes at certain times (e.g., car washing and lawn sprinkling during dry periods in summer). In a country enjoying a high standard of

[4] Gilbert F. White in preface to National Academy of Sciences — National Research Council, Committee on Water, <u>Alternatives in Water Management</u> (Washington DC: NAS-NRC, 1966).

[5] The terms were used by the US Congress, Senate Select Committee on National Water Resources, <u>Water Resources in the United States</u>, 86th Cong., 2nd Sess., Comm. Print No. 31 (Washington DC: GPO, March 1960), p. 22.

living the diminution of service by means of legal compulsion should be considered a short period palliative rather than a long-term solution. [6]

It is preferable to use price as an instrument of policy in such a way as to reduce those uses of residential water that the consumer values less than the cost to the community of providing the service. This approach leaves the consumer with the choice of exercising his right to buy more water at a price that reflects its cost; at the same time the management makes use of non-arbitrary criteria in attempting to allocate resources efficiently to the development of residential water supplies.

In appraising man's use of his environment in pursuit of a livelihood, it is interesting to examine whether resources are being allocated efficiently (i.e., in such a manner as to maximize socio-economic benefits[7] or to reach some other stated objective at minimum cost). The research strategy adopted in this study is to identify a set of variables, some of which may be subject to control (e.g., price), which may be expected to affect the level of residential water use and consequently the amount of investment required to provide this service. The level of residential water use may be considered as a target variable and those 'explanatory' variables that are subject to control may be considered as instrument variables. [8] This method of proceeding requires the formulation of hypothesized structural relationships among the component variables and then verifying the proposed model against a sample of observations. The results apply in varying degrees to different milieux but the experiment can be repeated and evaluated as often as required.

The issue is not only of academic interest. The rapidly growing urban population in all parts of the world requires heavy

[6]Emergency restrictions reduced water use in New York by 25% in 1965; see J. Hirshleifer, J.C. DeHaven, and J.W. Milliman, Water Supply: Economics, Technology, and Policy, 4th impression (Chicago: University of Chicago Press 1969), p. 371. These authors advocate pricing as a policy alternative but others argue that the influence of price, even if significant, is not very strong in magnitude and foresee political difficulties in raising prices. See US Congress, Senate Select Committee on National Water Resources, Future Water Requirements for Municipal Use, 86th Cong., 2nd Sess., Comm. Print No. 7 (Washington DC: GPO, January 1960), pp. 16-17; and S.J. Turnovsky, "The Demand for Water: Some Empirical Evidence on Consumer's Response to a Commodity Uncertain in Supply," Water Resources Research 5, no. 2 (1969).

[7]This is one of the stated aims of the Federal Council for Science and Technology, Committee on Water Resources Research, A Ten-Year Program of Federal Water Resources Research (Washington DC: GPO, 1966), p. 2.

[8]The terminology is adapted from Jan Tinbergen, Economic Policy: Principles and Design, 4th rev. printing (Amsterdam: North - Holland Publishing Co., 1967), chap. 1.

investment for residential water, which is generally considered as a component of the infra-structure required for healthy city growth. There were 500 million people suffering from water-borne diseases in the world in 1960.[9] These are mainly but not exclusively in the developing countries. The US Task Force on Environmental Health and Related Problems points out that "fifty million Americans drink water that does not meet Public Health Service drinking standards. Another forty-five million Americans drink water that has not been tested by the Public Health Service."[10] In many parts of rural and suburban Ontario septic tanks and shallow wells sunk on residential properties are still common, with the possibility of contamination of drinking water.[11]

The emphasis on water demand management rather than on the 'supply fix' makes it possible to improve the position of more communities with the investment resources that are actually available. This emphasis also reflects the high level of projected investments required in this field. For example, the resources required to close the gap between the existing level of water supply and needs in urban areas in seventy-five developing countries with an urban population of 336 million has been estimated at $5,834 million over the period 1961-75.[12]

The lack of concern for discovering tools to manage residential water demand stems from the assumption that demand for residential water use is not responsive to pricing policies. Many large cities in developing countries have adopted Western-style promotional pricing schedules.[13] The problem of finding adequate investment to develop municipal water supply is more acute in the developing world. On the other hand the urban population of North America is expected to double by the year 2000, and a largely

[9] J. Logan "The International Municipal Water Supply Program: A Health and Economic Appraisal," American Journal of Tropical Medicine and Hygiene 9, no. 5 (1960):469.

[10] US Task Force on Environmental Health and Related Problems, A Strategy for a Livable Environment (Washington DC: GPO, June 1967), p. 13.

[11] Evidence of the efficacy of a safe piped water supply in combatting water-borne disease is summarized by Dieterich and Henderson, Urban Water Supply Conditions in Seventy-Five Developing Countries, pp. 20-21.

[12] Ibid., p. 56, table 6.

[13] Harold R. Shipman, "Water Rate Structures in Latin America," Journal of the American Water Works Association (hereafter referred to as Journal AWWA) 59, no. 1 (1967):3-12 and "Comment" by P.P. Azpurua, Journal AWWA 60, no. 6 (1968):743-745 and by Shipman, 745-748. An increasing price block schedule adopted in Caracas is reported by P.P. Azpurua et al., "New Water Rates for Caracas," Journal AWWA 60, no. 7 (1968):774-780.

expanded water supply is a cornerstone of this development. The physical supply of water for urban purposes is adequate for the forseeable future and there is no crisis in this respect;[14] but there is an urgent need of planning the expansion of the urban water supply efficiently. The US Senate Select Committee on National Water Resources point out that:

Where cities do run out of water, the difficulty in most instances, up to now has been not so much a shortage of water as a shortage of vision ... what is required in most instances is not more water as such but more forethought as to future needs and possibilities and the willingness to finance preparation of plans and the provision of the additional waterworks needed to bring the water to the people. The water itself is usually available.[15]

Similarly, that a city never runs short of water may be due to consistent overinvestment by officials who want to play safe.

SCOPE OF THE STUDY

The study focuses on the identification of variables that affect the level of residential water use and the level of the related investment in water supply (Figure 1). Some of these variables are less amenable to policy manipulation by the municipal water supply management (e.g., level of evapotranspiration, size of family, income level, density of housing). Other factors (e.g., price) may be varied by management with a view to discouraging some uses of water. Figure 1 outlines schematically the structure of residential water use management. The factors affecting residential water use may be conveniently classified into three groups: those that vary from region to region, those that differ from one water utility to another and those that vary from household to household.

[14] "...it would appear that for municipal purposes we shall not be running out of water in a quantitative sense." US Congress, Senate Select Committee on National Water Resources, Future Water Requirements for Municipal Use, p. 18. This committee estimated municipal withdrawals at 17 Billion Gallons Daily in 1960, 29-33 BGD in 1980 and 43-56 BGD in 2000 out of total estimated withdrawals for all uses in 280 BGD in 1960 and 600 BGD in 1980 (and possibly 1000 BGD by 2000). Ibid., pp. 17-18.
In Ontario the estimated withdrawals for Public Supply in 1985 are 1.25 BGD out of a total 11.6 BGD and in 2000 the estimated figures are 2.5 and 21.1 BGD respectively. See A.K. Watt, "Adequacy of Ontario's Water Resources," The Canadian Mining and Metallurgical Bulletin 60, no. 664 (1967):920, table III; see also, Gilbert F. White, Strategies of American Water Management (Ann Arbor: University of Michigan Press, 1969), p. vii.
[15] Future Water Requirements for Municipal Use, p. 18.

Inter-regional differences in residential water use arising from cultural considerations are outside the scope of the study; the study area is assumed to be homogeneous in this respect. Supply considerations affect the quality, volume, and dependability of residential water but they do not affect the potential rate of use (i. e. , the potential demand). The climatic and soil conditions affect the rate of residential water use (particularly for lawn watering), and the effects of climatic conditions are discussed below; unless there are strong relief features, however, climatic conditions show little variation over relatively small areas such as the study area.

Of the household characteristics that affect water use, only income and size of household are discussed; density of housing is controlled for by limiting the discussion to single-unit dwellings. The consumer's preferences, tastes, and habits are assumed to vary in a random manner except for variations due to income, size of household, and type of housing.

The group of factors that vary from one water utility to another includes variables relating, among others, to quality of service. The quality of service is uniformly high in the study area and therefore these factors are excluded from the analysis. Waste reduction through mains leakage control, as well as the allocation of water of different quality to different types of uses (e. g. , treated effluent for industrial use), are variables affecting the quality and availability of water rather than its rate of use by the consumer. Regulation and exhortation are relevant to the study; they constitute a diminution of the service (viz. , water on tap at all times in virtually unlimited quantities). No restrictions were placed on water use in 1967 in those municipalities from which observations were drawn.

The main emphasis in the analysis is upon metering and pricing. On the one hand the present structure of water rate schedules appears to be a major weakness of urban water use management in the study area and on the other hand pricing is a flexible manage-ment tool that leaves the final choice to the consumer.

The target variables (i. e. , the average annual, average summer, average winter, maximum day, and peak-hour residential water use) are related to the factors mentioned above; these structural relationships may be stated in the form of an equation and the hypothesized equation, when fitted to sample data, yields coef-ficients which may be used in the application stage of the manage-ment exercise. The various measures of residential water

OBSERVABLE CONDITIONS

FACTORS AFFECTING RESIDENTIAL WATER USE

MACRO-SCALE OR INTER-REGIONAL

A. Physical conditions affecting demand
  1. Climatic and soil conditions
     i. Precipitation
     ii. Temperature
     iii. Water balance

B. Physical conditions affecting supply
  1. Climatic
     i. Precipitation
     ii. Temperature
     iii. Water balance
  2. Surface water
     i. Amount
     ii. Quality
     iii. Dependability of flow
  3. Ground water
     i. Amount
     ii. Quality
     iii. Depth of aquifer below surface
     iv. Dependability of flow
  4. Long distance water transfers
  5. Desalination

C. Financial, legal and political constraints on the exploitation of water supply

D. Cultural milieu (e.g., religion)

MUNICIPAL OR INTRA-REGIONAL

A. Quality of service
  1. Safety
  2. Pressure
  3. Taste
  4. Odour
  5. Colour
  6. Hardness
  7. Purity
  8. Dependability
  9. Individual connections versus street standpipe
  10. Single-tape versus multiple-tap connections
  11. Continuous or intermittent service

B. Policy variables
  1. Regulation/prohibition on day, time, amount, and purpose of water use
  2. Water rate structure
     i. Metering
     ii. Minimum bill
     iii. Amount of water allowed with the minimum bill
     iv. Frequency of billing
     v. Number, 'width', and level of price blocks
  3. Educating consumers
     i. Free inspection and repair of faulty plumbing
     ii. Exhortation

MICRO-SCALE OR HOUSEHOLD

A. Density of housing (apartments, single-unit)

B. Density of occupance (size of household)

C. Income level
  1. Lot size
  2. Number, type, frequency of use of water-using appliances

D. Legal, cultural constraints or incentives (e.g., by-laws, type of neighbourhood)

E. Consumer's preferences, habits, and tastes (e.g., preference for gardening, holidays during holidays during summer/winter)

F. Connection to public sewer

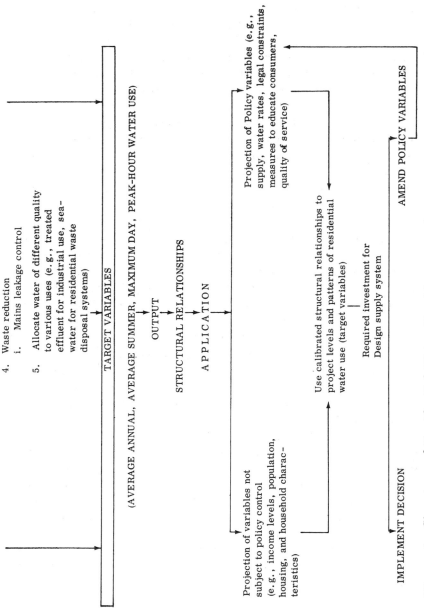

Figure 1. Structure of Residential Water Use Management

demand may be programmed if (a) the variables not subject to control are projected, (b) the variables subject to control are projected in accordance with some stated objective and (c) coefficients of water use are available from a fitted 'model'. If the outcome of the projection is not acceptable to management, the variables subject to control may be modified and a final acceptable projection of water use obtained through iteration.

The argument outlined above is worked out for a particular city-region and estimates of the coefficients are obtained from sample data. These computed coefficients are of a tentative rather than a definitive nature. However, a modus operandi is put forward together with a theoretical basis and some evidence from the Toronto region.

Some related topics are excluded from the main discussion. For example, the use of municipal water for non-residential purposes (e.g., commercial, industrial, public buildings, and fire protection) is outside the scope of this study. The emphasis on residential water use is justified because in many instances it accounts for a large proportion of the total municipal demand.[16] The residential demand also requires more investment per unit of volume delivered because the residential consumers are more widely distributed over the urban space than other users. Residential users also have a higher priority than other consumers and their demand is characterized by peak rates several times the average rate of use and which municipal water supply system management aims to meet.[17]

Municipal water supply and effluent disposal are related because municipalities discharge about 80 gallons of effluent per capita per day and in a highly urbanized region this vast quantity of water has to be treated in order to be reused by other users. A reduced level of water use will result in reduced expenditures on effluent control. However, pollution control and water quality are largely excluded from the discussion.[18]

The cost of supplying municipal water of potable standards are higher if the source is polluted. However, the cost of municipal

[16]Unofficial estimates in the study area varied from about one-half to three-fourths. The figure in Etobicoke, which maintains such records, was 58% in 1966 and 47% in 1967.

[17] AWWA Committee Report, "Basic Principles of a National Water Resources Policy," Journal AWWA 49, no. 7 (1957):831; and A. Wolman, "Providing Reasonable Water Service," Journal AWWA 47, no. 1 (1955).

[18] A useful recent guide to the topic is P. H. McGauhey, Engineering Management of Water Quality (New York: McGraw-Hill, 1968); see also, Sigurd Grava, Urban Planning Aspects of Water Pollution Control (New York: Columbia University Press, 1969).

water treatment is relatively low and the increase in such costs due to poor water quality does not justify very high levels of waste effluent treatment except in cases of extremely toxic or evil-tasting substances.[19]

ORGANIZATION OF THE STUDY
AND SOME WIDER IMPLICATIONS

The development of urban water supplies is discussed in the next section and some features of municipal water supply in the Toronto region are described in the final section of this chapter.

Previous findings on factors affecting residential water demand and the possibilities of forecasting and planning urban water supply and demand are the main topics of Chapter II. The survey of previous theoretical and empirical work in this field suggests a research strategy: the hypothesized structural relationships of residential water use are presented in Chapter III which also includes the fitting of multiple regression equations to sample data.

The statistics derived from the fitted equations are applied to selected target variables in Chapter IV. The cost structure and pricing methods of the industry are discussed first; water rates are the most suitable variables available to reduce the level of residential water use at peak periods without resorting to legal restrictions or curtailing supply by reducing pressure; each proposed alternative is considered first in theoretical terms and then an attempt is made to 'measure' its effect given the coefficients obtained from the fitted equations. The final chapter summarizes the findings and makes some suggestions for improving our understanding of this topic through further research and systematic data collection.

The use and misuse of resources by man and the study of the inter-relationship of phenomena over space is an attractive research theme to economic and resources geographers.[20]

---

[19] A. V. Kneese, _Economics and the Quality of the Environment: Some Empirical Experiences_, Resources for the Future Reprint No. 71 (Washington, DC: RFF, 1968), p. 175.

[20] An important paper in the development of this school is Harlan H. Barrows, "Geography as Human Ecology," _Annals Association of American Geographers_ 13, no. 1 (1923):1-14.

Although the development of water supply is a basic requirement for urban growth, the understanding of the determinates of water demand has been given less attention than the ways of increasing supply.

The research reported here explores policy alternatives to the projected investment of scarce public resources in municipal water supply systems; it is found that this new expenditure in municipal water works could be postponed or reduced without serious curtailment of the service. Therefore, a better understanding of residential water demand could be important for planning and managing urban water supply systems.

This stance may be adopted with respect to other — and more costly — public services (e.g., education, highways, recreation areas) that are characterized by (1) a need to cope with demand peaks of relatively short duration so that part of the system capacity is idle most of the time; (2) an imperfect pricing mechanism so that the marginal cost of use to the individual is less than the marginal cost of supplying the service at public expense; (3) general agreement that the service is basic to the quality of urban life and essential to the well-being of the community and the individual.

Diminishing marginal utility applies to these other public services too and if the pricing mechanism is applicable to residential water demand management, the same policy tools may be applicable to the demand of other public services such as higher education, highway transport, and recreation. Besides the argument for the efficient allocation of scarce public funds, there is also the argument for social equity. The demand for recreation, higher education, highways, water, and all other goods and services is correlated positively with income, generally speaking. Therefore the pricing of these public services below the marginal cost of producing them is equivalent to a hidden subsidy to those who can afford to make use of such public services.

The rocketing expenditures on higher education, highway transport, recreation facilities, water supply, and other public services raise the question whether the demand for some of these services cannot be 'controlled' or managed in such a way that those who make the most intensive demand (e.g., those who use the automobile for all trips, those who demand secluded parks rather than just parks for swimming, camping, etc., those firms and institutions which demand staff with a total of 21 years schooling and training) pay an appropriate marginal price for the public service. One major difficulty is that the cost to the individual for public services is not easily measured and therefore it would be difficult to impute prices to the demand for such services. The problem of demand management is more amenable to critical examination with respect to residential water supply where there is a pricing mechanism which may be observed.

HISTORICAL PERSPECTIVE

The story of municipal water supply goes back to biblical and classical times.[21] After a more or less complete break during the Middle Ages, modern municipal waterworks developed during the industrial revolution and its attendant urban growth. By 1800 some European and North American cities (e.g., Paris, London, Boston, and New York) had modern waterworks including the use of steam engines to pump water and a distribution system leading into private homes but the service was on an intermittent basis until the late 1800's.[22]

The main impetus to the building and expansion of municipal waterworks was provided by the need to control fires, and the conviction that polluted drinking water was the source of disease such as cholera and typhoid fever.[23] Filtration on a large scale was inaugurated in London in 1829 and by the 1870's almost all large cities in Europe and North America had a filtered water supply. The dramatic decrease of certain diseases following the

[21]The earliest detailed description of a municipal water supply system was written by a professional engineer who was the superintendent of Rome's waterworks. See Frontinus, De Aquis Urbis Romae, trans. Clemens Herschel (Boston: Dana Estes & Co., 1899). The Athens waterworks are described in American School of Classical Studies at Athens, Waterworks in the Athenean Agora (Princeton, N.J.: 1968).

[22]For a useful historical sketch and a bibliography on early municipal waterworks see F.E. Turneare and H.L. Russell, Public Water Supplies: Requirements, Resources and the Construction of Works, 4th ed. (New York: Wiley, 1940), chap. I.

[23]These themes run through the standard work by N.M. Blake, Water for the Cities: A History of the Urban Water Supply Problem in the U.S. (Syracuse: Syracuse University Press, 1956). The relationship between the provision of public water supplies and health is discussed in a paper by A. Wolman and H.M. Bosch, "U.S. Water Supply Lessons Applicable to Developing Countries," Water, Health and Society, selected papers by Abel Wolman edited by Gilbert F. White (Bloomington, Ind.: Indiana University Press, 1969), pp. 219-233; and in T.R. Lee, Residential Water Demand and Economic Development, University of Toronto Department of Geography Research Publications No. 2 (Toronto: University of Toronto Press, 1969).

introduction of a piped—and chlorinated—water supply may be illustrated from Toronto's experience. In the city of Toronto the death rate from typhoid fever was 0.41 per 1000 in 1910 and 0.073 in 1921 and 0.007 in 1940.[24] The improvement may be ascribed to the provision of filtered and chlorinated water after 1912.[25]

The salient features of the modern development of municipal waterworks may be noted because they are relevant to the current modes of management.

1. The growth of urban water systems was a response to the discovery that a purified, piped water supply reduced the incidence of certain water-borne diseases and also made possible the control of losses caused by fires in the densely built up areas of cities. A copious and cheap supply of water was also intended to help make the city more attractive by encouraging lawn sprinkling, street washing, and flowing fountains in gardens. All these 'good' uses lend support to the prevalent notion that a municipal water supply is beneficial not only to individuals but also to the community as a whole; the minimum bill for residential water may be interpreted as a contribution by the householder toward the upkeep and expansion of an essential public service. In addition it follows that the use of water ought to be encouraged rather than sold like other goods.

2. Urban waterworks require large-scale investment and are suited to be managed by one 'firm'. Since the commodity is so basic to the community's well-being, the ownership and operation of city waterworks have in most places been taken over by the municipality. Where private companies still survive they invariably enjoy a monopoly position and consequently there has been no significant conflict between public and private interests for the ownership of urban water supply plant in recent times.[26]

3. Charges were initially low in order to encourage residents to install the necessary plumbing fixtures. On the other hand the

[24]A. E. Berry, "Environmental Pollution and its Control in Canada: A Historical Perspective," in Pollution and Our Environment, Background Paper A-1, Canadian Council of Resources Ministers Conference, Montreal 1966 (Ottawa: Queen's Printer, 1967), p. 3.

[25]Two filtration plants were built on Centre Island in 1912 and 1917 with a combined capacity of 100 MGD. See Toronto, Board of Review, Board of Review on Sewage Treatment for the City of Toronto; Majority and Minority Report (Toronto, 1939), p. 13. Chlorination of the city water began on March 19, 1910. See A. E. Berry, "A Tribute to Norman Howard Joseph," Journal AWWA 56, no. 10 (1964):1371.

[26]J. W. Milliman, "Policy Horizons for Future Urban Water Supply," Land Economics 39, no. 2 (1963):110.

waterworks have traditionally been expected to pay their way —
unlike roads, schools, courts of law, museums, and libraries.
Therefore, the charges were designed to capture some of the
consumer's surplus with respect to the all-important first units
of consumption by means of minimum bills, meter service charges,
or other types of demand charges.

4. Municipal water systems have expanded as if unlimited
quantities of treated water, on tap at all times and at competitive
prices, are essential to the growth of a city-region. Even if the
water is not a direct source of revenue to the municipality, it is
an indirect source of assessment taxes. There is also an element
of pride among municipal officials in being able to meet the
citizens' demand for water.

5. It is generally thought that demand for residential water is
inelastic with respect to price changes because it has few competing substitutes, it is a small item in the family budget, and it
is an essential item to the household. As a result water use has
been curtailed by means of restrictions (e.g., lowering pressure
or legal prohibition), or exhortation rather than by increasing
prices up to the level of marginal costs. Since legal and physical
restrictions are unpopular, projected increases in 'demand' or
'requirements' result in additions to the waterworks.

There is enough substance in the logic underlying the propositions stated above to make them survive as plausible truths
among many decision-makers in the water utility industry. All
five traditionally-held views have helped to raise the standard of
service in terms of water quality, lack of shortages, adequate
pressure, and dependability of supply. These high standards
are easily justified for a service that pays for itself, is nearly
always run by the community itself and not by a private firm
for profit, is 'essential' to health and economic growth and is
cheap enough to be within reach of everyone. The traditions
outlined above testify to the high professional standards in the
water utility industry; in the field of development philosophy and
pricing policies however the industry has a long way to go.
Management and development practices that were justified in the
infancy stage of development have to be re-examined in the light
of changing social and economic conditions.

The waterworks industry may well be entering another phase
of its development. The marginal cost of urban water supply
(particularly for peak periods of demand) is making it an 'eco-

nomic good' while its per capita consumption at an increasing
rate is taking residential water out of the realm of 'social goods'.
The size of required investment is increasing and the competition
for public funds is not likely to slacken; at the same time fresh,
clean water is not the abundant resource it used to be. In this
new phase of municipal water supply development characterized
by increasing demands and costs, it will be important to examine
and understand better the 'requirement' for residential water and
to enquire whether the optimal allocation of resources through
(marginal) pricing is relevant to planning and management in this
field. Some evidence from previous studies is presented in the
next chapter; in the next section some features of the municipal
water supply in the Toronto Region are described in the light of
the discussion so far.

THE STUDY AREA

The water resources of Ontario are abundant in relation to projected levels of water use. Precipitation is estimated at about
486 billion gallons daily of which just over two-thirds falls in
river basins draining to Hudson Bay and one-third into basins
draining into the Great Lakes and St. Lawrence River. Evapo-
transpiration losses reduce this to 187 BGD. Of this amount,
the sustained yield is about one-half or 94 BGD which is available
as surface run-off and ground water. In addition, Ontario has
access to four of the Great Lakes which constitute about 20 per
cent of the world's stored supply of fresh water.[27]
The projected total demand for water is shown in Table 1.
The forecast assumes a population of 14 million by the year 2000
and a doubling in per capita water use. A comparison of projected
water use in Ontario and the potential supply "suggests that the
supply of water available to meet future needs is most adequate."[28]
However, the projected demand makes it clear that expenditures

[27]Watt, "Adequacy of Ontario's Water Resources," pp. 918-919; see also, "Our Water
Resources - What are They?" in Proceedings, Conference on Water Resources Manage-
ment, Toronto, 1966 (Toronto: Conservation Council of Ontario, 1966).
[28]Ibid.; see also J. G. Warnock, "Our Water Needs — What Will They Be?" in Proceedings,
Conference on Water Resources Management, Toronto, 1966 (Toronto: Conservation
Council of Ontario, 1966).

TABLE 1. FORECAST WATER USE IN ONTARIO
(Billion Gallons Daily)

	1966	1985	2000
Private Domestic	0.1	0.1	0.1
Irrigation	2.0	4.0	6.0
Public Supply	0.8	1.25	2.5
Industrial	4.0	6.25	12.5
	6.9	11.60	21.1

SOURCE: A. K. Watt, "Adequacy of Ontario's Water
Resources," The Canadian Mining and Metallurgical
Bulletin 60, no. 664 (1967):920, table III.

on both water supply and water pollution control will be very great.
While an emergency in potential water supply is very unlikely there
is some urgency in planning the urban water supply in such a way
that overbuilding and overinvestment are avoided without impairment
of the service. The Chairman of the Ontario Water Resources
Commission and some of his officials suggest that "the total
national costs associated with water supply and pollution control
are large and every effort must be made to minimize the effects
on the economy."[29]

Investment expenditures are likely to increase at a faster rate
than population growth partly because per capita investment requirements will increase as (1) the rate of per capita water use
increases particularly during peak demand periods and (2) the
sources of supply nearby are exhausted and sources some distance
away are developed. A considerable part of the cost of producing
residential water is the interest and repayment of debentures; in
addition the decision to enlarge the water supply system is related
to investment cost since there are indivisibilities in capital requirements and investment in residential water supply is almost
wholly irreversible.

The investment requirements per capita for waterworks is
already considerable. In the United States, where conditions may
be expected to be comparable to those obtaining in Canada, new

[29] J. A. Vance, K. E. Symons, and D. A. McTavish, "The Diverse Effects of Water Pollution on the Economy: Domestic and Municipal Water Use," in Pollution and Our Environment, Background Paper A 4-1-5, Canadian Council of Resource Ministers Conference,
Montreal, 1966 (Ottawa: Queen's Printer, 1967), p. 2.

investment in municipal waterworks per additional person served
was \$275.70 during 1956-65.[30] The per capita capital require-
ments of the water utility industry in the US are estimated as
follows (in 1965 dollars):

1966	1970	1975
\$310	\$350	\$400

The estimate includes provision to correct system deficiencies
e.g., quality problems of taste, odour, hardness, and corro-
siveness, and also capacity deficiencies; these deficiencies are
assumed to be corrected over a 15-year period. The total
expenditure during the decade is estimated at \$24 billion.[31] In
Ontario new investment in municipal waterworks during 1957-67
was \$446.2 million (actual dollars) or about \$40 million per
annum[32] (Table 2). Investment per additional resident in the
province during the decade 1958-67 was \$305.29 (1965 dollars)
or \$337.05 (1967 dollars). This is higher than investment in the
United States. Moreover the capital expenditure on waterworks in
Ontario is increasing at a higher rate than the population: the
population increase in Ontario has been 2 per cent since 1956 but
the waterworks capital expenditure in constant dollars has been
increasing by 3.5 per cent per annum.[33]

In the past, solutions to the increased demand for urban water
were, relatively speaking, small in scale since every municipality
had access to a nearby source of water (wells, river, or lake).
As the urban centres grow the municipalities have to go farther
in order to exploit new sources of supply except those municipalities which already take their water from a large nearby lake
(e.g., Toronto and Hamilton from Lake Ontario).[34] The problems

---

[30] The total new investment during the decade was \$10,200 million. See AWWA Staff
Report, "The Water Utility Industry in the United States," Journal AWWA 58, no. 7
(1966):772, table 8.

[31] Ibid., p. 779 and table 18 (p. 781).

[32] Besides the \$446.2 million in waterworks expenditures, nearly another 1 billion dollars
were approved for sewage works. This may be compared with the \$70 million invested in
the Winnipeg Floodway and the \$400 million for the projected Fraser River Project, both
of which have been the subject of investigation to assess economic benefits. Figures for
investment in waterworks in Ontario are compiled by the Sanitary Engineering Division,
Ontario Water Resources Commission.

[33] Unpublished report, Sanitary Engineering Division, Ontario Water Resources Commission, 1968.

[34] Toronto's first municipal water supply in the early 19th century was a well dug in the
yard of the Jail, near Toronto Street; see City of Toronto, Department of Public Works,
Report on City of Toronto Water Distribution System (Toronto, April 1968), p. 2. The
move from well to surface water as a source of supply for the public waterworks is typical
in Southern Ontario.

TABLE 2.   CAPITAL EXPENDITURES ON WATERWORKS 1957-67

Year	Southam Construction Index (1949/50= 100)	Expenditures Actual Dollars	Constant Value (1965) Dollars	Ontario Population (000's)
1957 (9 months)	141.5	29,511,711.60	39,898,165.58	5636.0
1958	145.8	39,224,132.67	51,464,589.95	5821.0
1959	152.0	42,258,878.41	53,185,022.63	5969.0
1960	157.6	29,892,077.75	36,283,975.09	6111.0
1961	157.9	28,931,973.56	35,051,846.37	6236.1
1962	161.9	51,643,155.53	61,021,220.83	6351.0
1963	174.2	33,456,802.65	36,741,023.81	6481.0
1964	182.1	54,961,748.83	57,778,509.34	6631.0
1965	191.3	44,526,956.69	44,526,956.69	6788.0
1966	198.6	46,321,229.04	44,618,585.67	6960.9
1967	211.2	45,519,615.79	41,230,598.96	7149.0

SOURCES: Population figures fromProvince of Ontario Registrar-General, Vital
Statistics for 1967 (Toronto: Queen's Printer, 1969); Capital Expenditures figures
from Sanitary Engineering Division, Ontario Water Resources Commission;
Southam Construction Index from Engineering and Contract Record (Southam
Business Publications Ltd., Don Mills, Ontario) 81, no. 12 (1968):33.

and tentative solutions are outlined by a former General Manager
of the Ontario Water Resources Commission:

Now, as the inland areas of the province increase in population and development, the problem of adequate water supplies takes on new significance.
Inland communities in a large part of southern Ontario have serious water
problems, including quantity, quality and pollution control. Costs for the
development and transportation of water have increased sharply and high
interest rates have added to them. Where water must be piped long distances, it is an advantage to deal with this problem on a regional basis
rather than for a number of municipalities to act separately.[35]

The solution to increasing demand for urban water is likely
to require larger outlays of capital because of increasing distances
(and sometimes the change from well to lake water supply); the
management of municipal water supply systems is also likely to
become more complex. The level of water quality deterioration

[35]A.E. Berry, "Ontario Water Resources Act," Journal AWWA 50, no. 9 (1958):1127.

21

and the proximity of urban centres have made regional solutions attractive. This tendency is reinforced by the move favouring regional governments in Ontario. In 1954 Metropolitan Toronto took over responsibility for water supply and mains distribution for the thirteen constituent municipalities.[36] In 1956 the Ontario Water Resources Commission was established[37] and there are now 25 regional pipelines under construction.[38] Some of these pipelines have since been built or are under construction e.g., Lake Huron-London, Lake Erie-St. Thomas, South Peel County (Mississauga, Brampton, and Chinguacousy). Partly as a result of the increasing capital outlays required for these pipelines, there has been a trend for part of the water supply management function to be taken over by a higher level of government (Provincial, regional, or metropolitan).

In such cases the higher level of government takes over the supply functions of the urban water management; this is a reflection of the heavy and indivisible investment requirements and the possibility of obtaining economies of scale and easier financial terms. The municipal government unit still manages the local distribution facilities, pricing schedules, billing, and so on. This fragmentation of the management functions of metropolitan water supply systems, in some instances, may exclude some policy alternatives; for example, if one municipality in a metropolitan unit tries to reduce peak demands and the others do not, the benefits would be spread over the whole group of municipalities. If the whole supply unit is considered the savings in investment would be more direct and obvious.

The southern part of central Ontario, where metropolitan growth is most evident, faces the problem of higher investment in municipal water supply and water pollution control, as well as the challenge of urban planning and regional government. The study area in that part of south-central Ontario that is

---

[36]It has been suggested that the "supply of water was one of the major reasons for the advent of 'Metro.'" The transfer of responsibility for water supply and sewage to the Metro authority gave the whole urban area access to the lake, the obvious source of supply and terminal for the disposal of treated sewage; before 1954 those municipalities not having lake frontage depended on wells or water purchases from neighbouring municipalities. See Frank J. Horgan, "Progress Report on the Metropolitan Toronto Supply," Journal AWWA 56, no. 10 (1964):1297-1302; for a similar view see L.B. Allan, "Water and Sewage Works in Metropolitan Toronto," The Municipal Utilities Magazine 92, no. 4 (1954):54.

[37]The objectives and instruments of this Act are summarized by Berry, "Ontario Water Resources Act," pp. 1127-1131 and "Discussion," pp. 1131-1135.

[38]H.J. McGonigal, "Economic Aspects of Environmental Quality for Ontario," Ontario Economic Review 8, no. 2 (1970):7.

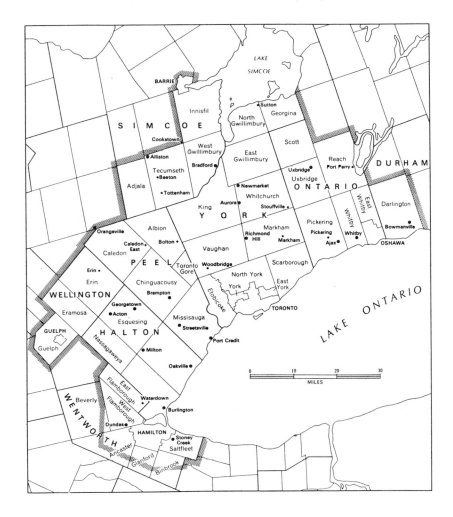

Figure 2. The Study Area

enclosed by a line joining Hamilton, Guelph, Barrie, and Oshawa
(Figure 2). It is a fast growing urban area which is attracting
some attention from the provincial government because of its
relative importance to Ontario in terms of population, employment,
and pressing problems that are characteristic of coalescing
conurbations, e.g., pollution of air and water, urban sprawl, the
need to expand urban water supply facilities and other public
services, and the problems of formulating suitable forms of
regional planning and government. This area was the subject of

23

TABLE 3. PROJECTED INVESTMENT IN MUNICIPAL WATERWORKS
IN THE STUDY AREA, 1964-2000 (1967 Dollar Values)

Year	Population (Millions)	Increase in Population (Millions)		Projected Investment (Million Dollars)
1964	2.8	--		--
1980	4.0	1964-1980	1.2	404.4
2000	6.4	1980-2000	3.6	1213.2

a study known as the Metropolitan Toronto and Region Transportation Study (MTARTS); among other 'suggestions' in this report was a pipeline diverting water from Georgian Bay southwards into the basins draining to Lake Ontario.[39] Water supply is considered to be a critical factor in urban development in this region and this suggestion "will have the effect of removing present restraints on urban growth arising out of limited or dubious water supply."[40]

The projected investment in municipal waterworks in the study area is shown in Table 3. The population projection is adopted from the Metropolitan Toronto and Region Transportation Study[41] and it is assumed that an investment of $337 (1967 dollars) per new resident is required; this figure is based on the average figures for Ontario quoted above.

An expenditure of about $1.2 billion on municipal waterworks in the most urbanized part of Ontario over the next three decades is not excessive, especially when compared to other items in the public budget. It is substantial enough, however, to warrant a closer look at alternative management practices which may reduce the investment requirements without impairing the service.

The region is adopted as a frame of reference for examining the possibilities of residential water demand management because there is some variety among the municipal water supply systems

[39]MTARTS, Choices for a Growing Region, Second Report (Toronto: Department of Municipal Affairs, Community Planning Branch, November 1967), p. 52 and map 16, p. 53. The pipeline is envisaged as serving Barrie, Alliston, Orangeville, and Guelph each with a population of 250,000.

[40]Ibid., p. 52. See also G.H. Kay, A.R. Townshend, and K.F. Lethbridge, "Water Pollution Control and Regional Planning — The Grand River Watershed," in Pollution and Our Environment, Background Paper B 17-1-3, Canadian Council of Resource Ministers Conference, Montreal, 1966, vol. 2 (Ottawa: Queen's Printer, 1967).

[41]Ibid., p. 9, table 1.

with regard to sources of supply and management practices (e. g. , price schedules). At the same time it is sufficiently homogeneous in terms of inter-regional variables such as climatic conditions.[42] Although it includes one-third of the population of Ontario the study area is relatively small in areal extent: it is 3, 190 square miles; the population in 1964 was 2. 73 million, an increase of 800, 000 or 36 per cent from 1956.[43]

The research strategy required a sample from a region which offered variety in management practices, particularly differences in the level of the marginal price for residential water. A sample from a wider area would have necessitated the inclusion of variables such as frequency of precipitation during summer, length of growing season, and so on. Such variables, however relevant to an understanding of residential water use, are not controllable by the municipal water utility management. Thus, a compact study area which included sufficient variety in management practices was selected;[44] in fact most of the sampling was done within a smaller core area, the Metropolitan Toronto Planning Area (MTPA),[45] but the pricing range in this core area was limited.

[42]The study area has a temperature in July of 66° to 70°F; the average May to September rainfall is greater than 12" and less than 16"; the start of the growing season (temperature above 42°F) is later than April 10 and prior to April 20; the end of the growing season varies between October 31 and November 5 and the average annual actual evapo-transpiration (4" storage) is about 20.5". See L. J. Chapman and D. M. Brown, The Climates of Canada for Agriculture, Canada Land Inventory Report No. 3 (Ottawa: Queen's Printer for the Canada Department of Forestry and Rural Development, 1966).
The above statistics indicate that the variation over the study area of climatic conditions is small. In a study discussed in the next chapter, Haver and Winter included a weather variable (number of days in June, July, and August with a rainfall of 0. 01 inches or more) in their estimating equation of water use/capita/day. This variable did not "improve the relationship significantly. " Their study included 13 cities from all over Ontario (e. g. , Kingston, London, Peterborough, Richmond Hill, Chatham, and Waterloo) but it included the total water use in the municipalities.

[43]MTARTS, Growth and Travel: Past and Present, First Report (Toronto: Department of Municipal Affairs, Community Planning Branch, April 1966), p. 80, table 1.

[44]In setting up a hypothetical relationship such as $Y = f(X_1, X_2)$ it is assumed that excluded variables are independent of the ones that are included as explanatory variables. When the level of water use in two or more regions is compared, the relevant climatic variables (e. g.', frequency of rainfall during the summer months) should be included in the equation since the excluded climatic variables may be systematically related to management practices (e. g. , high prices and marked water deficiency). The choice of a compact area for sampling purposes reduces the climatic variations.

[45]The MTPA covers one-fifth of the area included in the MTARTS and includes over two-thirds of the population. The MTPA embraces the Municipality of Metropolitan Toronto, the Townships of Mississauga, Toronto Gore, Vaughan, Markham, and Pickering; the towns of Port Credit, Streetsville, Richmond Hill, and Ajax; the villages of Woodbridge, Markham, Stouffville, and Pickering.

Demand for water provides the initial impetus to the building of larger and more complex urban water supply systems; the alternatives open to management on the demand side are relevant to urban water resources planning in this region which is passing through a phase of growth and a phase of change in the sources of supply for some municipalities. While detailed applications are not attempted — due to lack of planning data readily available — the study area offers ample scope for specific applications of the analysis attempted in this study.

Table 5 and 6 in the Appendix to this chapter provide background information about expenditures, consumption, sources of supply, treatment, metering, and pricing. Some of these data are summarized in Table 4 below. Twenty-two out of the 53

TABLE 4. MUNICIPAL WATER SUPPLY SYSTEMS
IN METROPOLITAN TORONTO AND REGION

Source	Size Category	No. of Municipalities		1967 Consumption (Million Gallons)		1967 Population Served	
		Metered	Non-Metered	Metered	Non-Metered	Metered	Non-Metered
Lake	A	8	3	43,356.9	65,220.5	1,493,891	1,057,324
	B	2	4	476.5	707.6	22,892	38,776
Wells	A	1	1	950.1	365.7	25,055	19,468
	B	10*	12*	1,681.5	1,030.9	84,351	39,098
Rivers	A	-	-	-	-	-	-
	B	-	1	-	166.2	-	5,867
Springs	A	-	-	-	-	-	-
	B	-	1*	-	-	-	994
Other (e.g., mixed, purchased)	A	2	-	2,540.0	-	87,523	-
	B	5*	3	706.3	507.4	49,679	12,891
TOTALS		28	25	49,711.3	67,998.3	1,763,351	1,174,418

NOTES: A stands for population greater than 19,000
B stands for population less than 19,000
* The number of municipalities for which data is included in the columns for consumption is one less than the number marked *.

SOURCE: Ontario Department of Municipal Affairs, 1967 Annual Report of Municipal Statistics (Toronto: Queen's Printer, 1968).

municipalities have populations less than 19,000 and are served by wells; two other municipalities with a population greater than 19,000 are also served by wells (Barrie and Richmond Hill). All these municipalities may be expected to outgrow their present source of relatively cheap water (i.e., well water). The transfer to other sources of water supply such as a pipeline from a large lake would require considerable capital expenditures which could be postponed or reduced if the demand for water were reduced through a rational pricing schedule.

About two-fifths of the population in the study area have non-metered residential water consumption; nearly half of the municipalities have a non-metered water supply for residential customers, including the two largest (Toronto City and Hamilton). It will later be argued that the level of water use by non-metered residential consumers is well above that used in metered residences. If metering is introduced in the non-metered municipalities, the reduction in water use could offset in part the growing requirements over time, thus making it possible to postpone capital expenditures. Eighteen municipalities drew their water from a source other than a large lake and for these municipalities the impact of marginal pricing may delay for some time the necessity of changing to a less accessible source of water supply. It is clear that residential water demand management may have considerable pay-offs in the efficient allocation of resources for this essential service in the study area.

TABLE 5. MUNICIPAL WATER UTILITIES IN METROPOLITAN TORONTO AND REGION: EXPENDITURES, SOURCES, TREATMENT, METERING STATUS, AND SIZE, 1967.

	(1) Total Expenditure $	(2) Population Served	(3) Expenditure/ Capita $	(4) Source of Supply	(5) Treatment C & F	(6) Consumption (Million Gallons)	(7) Metered (M) Non-metered (U) Partly metered (M?)
City of Toronto	12,510,794	685,313	18.27	P	-	39,560.7	U
Etobicoke	3,794,517	266,548	14.24	P	-	9,839.3	M
Scarborough	2,994,012	275,632	10.86	P	-	7,971.2	M
York	1,526,522	141,562	10.78	P	-	4,397.6	M
East York	1,050,027	97,555	10.76	P	-	3,342.4	M
North York	4,907,295	411,500	11.92	P	-	11,428.0	M
TOTAL	26,174,789	1,878,110	13.94	L	C & F	(77,397.7)	-
Barrie	236,592	25,035	9.45	W	C	950.1	M
Guelph	502,285	51,873	9.68	S & W	C	1,550.0	M
Hamilton	4,790,709	291,536	16.43	L	C & F	22,292.1	U
Oshawa	1,116,440	80,475	13.87	L	C & F	3,367.7	M(?)
Acton	62,098	4,460	13.92	W	C	74.6	M
Ajax	172,483	12,298	14.02	L	C & F	516.6	M(?)
Alliston	89,380	3,085	28.97	L & R	C & F	165.7	U
Aurora	115,635	11,292	10.24	W	-	-	U
Bowmanville	83,880	8,358	10.04	L & S	C	-	M
Bradford	43,294	2,700	16.03	W	-	108.5	U
Brampton	627,428	35,650	17.60	W & P	C & F	990.0	M
Burlington	801,776	65,376	12.26	L	C & F	1,771.2	M
Dundas	248,389	15,752	15.77	P	C & F	293.3	M
Georgetown	194,178	12,798	15.17	W	C	400.0	M
Milton	94,194	7,270	12.96	W	-	461.0	U
Newmarket	120,738	-	-	-	-	-	M
Oakville	925,822	50,718	18.25	L	C & F	1,314.2	M
Orangeville	53,663	5,847	9.18	S & W	-	228.0	U
Port Credit	84,098	7,892	10.65	L	C & F	219.8	M
Richmond Hill	236,683	19,468	12.16	W	F	365.7	U
Stoney Creek	111,119	7,758	14.66	L	C	204.6	U

				Source	Treatment		
Streetsville	57,035	5,867	9.72	R	C & F	166.2	M(?)
Uxbridge	23,979	3,026	7.92	W	-	68.0	U
Whitby	343,046	15,090	22.73	L	C & F	411.2	M(?)
Beeton	8,402	994	8.45	S	C	-	U
Bolton	30,443	2,225	13.68	W	-	75.0	U
Caledon East	10,635	695	15.30	W	-	6.0	U
Cookstown	15,532	705	22.03	W	-	6.7	U
Erin	10,868	1,068	10.18	W	C	26.0	U
Markham	82,830	8,261	10.03	W	?	156.9	M
Pickering	38,110	1,943	19.61	P	-	30.5	M
Port Perry	49,214	3,023	16.28	W	C	52.3	U
Stouffville	31,221	3,959	7.87	S & W	C	113.7	U
Sutton	46,781	3,800	12.31	L	C	40.2	U
Tottenham	6,262	783	8.00	W	-	15.0	U
Waterdown	33,895	2,425	14.00	W	-	31.0	M
Woodbridge	36,972	2,944	12.56	W	-	185.0	U
Ancaster	179,581	10,500	17.10	W & P	?	130.6	M
Chingacousy	121,641	37,500	9.73	W	-	401.9	M
Gwillimbury E.	27,071	4,563	5.93	W	C	43.6	M
King	60,411	3,367	17.94	W	-	27.4	U
Markham Township	347,447	14,880	23.35	W	C	211.0	M
Pickering Township	159,749	15,000	10.65	L	C & F	256.7	M
Saltfleet	327,348	13,126	24.94	P	-	259.1	M
Toronto	1,926,494	105,000	18.35	L	C & F	3,293.0	M
Vaughan	429,100	12,575	34.12	W	C & F	350.0	M
Whitby	18,622	1,889	9.86	W	C	12.5	M
Whitchurch	15,689	?	9.86	?	?	?	?

NOTES: Source of water supply:-
P Purchased
L Lake
W Wells
R Rivers
S Spring

Treatment:-
C Chlorination
F Filtration

SOURCE: Ontario Department of Municipal Affairs, 1967 Annual Report of Municipal Statistics.

TABLE 6. METERED RESIDENTIAL WATER USE FOR
MUNICIPALITIES INCLUDED IN THE STUDY AREA[1]

Municipality	Population[2]	Price of Water[3] Cents/1000 Gallons	Billing Period[3] (Months)	Allowance with Minimum Bill per Quarter (Gallons)
Metropolitan Toronto				
Etobicoke	266,456	39	3	6,000
Scarborough	275,632	(change in 1967)	2	6,000
York	141,562	39	3	5,500
East York	97,500	44	3	3,240
North York	411,500	(change in 1967)	4	6,000
Cities				
Barrie	25,035	36	2	15,000
Guelph	51,873	36	2	6,000
Oshawa	80,475	38.4	2	-
Towns				
Acton	4,429	40.5	2	-
Ajax	10,355	34	3	12,000
Bowmanville	8,332	36	2	5,625
Brampton	14,727	45	2	6,000
Burlington	121,610	30	2	6,000
Dundas	15,752	-	-	-
Georgetown	12,623	30	2	7,500
Newmarket	10,000	80	2	1,890
Oakville	50,718	50	2	9,000
Port Credit	7,892	47.5	2	-
Richmond Hill	19,428	41.5[5]	3	-
Streetsville	5,867	-[5]	-	-
Whitby[4]	15,000	51.3, 36	2	-
Villages				
Markham	8,161	36	2	-
Pickering	1,943	71.25	2	6,000
Waterdown	2,425	50	2	3,000
Townships				
Ancaster	10,500	40	4	3,000
Gwillimbury East	4,563	32	3	-
Chinguacousy	7,276	45	3	10,000
Markham[4]	14,800	100, 40	2	7,500
Mississauga	105,000	45	3	6,000
Pickering	15,000	75 (with sewerage charge)		
		50	2	10,000
Saltfleet	13,126	57	3	6,000
Vaughan	12,200	36	3	9,000

NOTES:
1. Metropolitan Toronto and Region Transportation Study (MTARTS), <u>Growth and Travel: Past and Present</u>, First Report (Toronto: Government of Ontario, 1966). The following municipalities are unmetered: Toronto City, Hamilton, Alliston, Aurora, Bradford, Milton, Orangeville, Stoney Creek, Uxbridge and the villages of Beeton, Bolton, Caledon East, Cookstown, Erin, Port Perry, Sutton, and Tottenham. In addition townships with a population of 5,000 are usually omitted (see note 2).
2. 1967 population figures are from Ontario Department of Municipal Affairs, <u>1967 Annual Report of Municipal Statistics</u> (Toronto: Queen's Printer, 1968, Waterworks Section) p. 186 ff. Townships with a population of 5,000 or less are not listed in the waterworks section and are therefore omitted from this list.
3. Prices are shown net of discounts and are taken from Stanton Pipes (Ltd.) "1967 5th Annual Survey of Municipal Water Rates of Ontario, " <u>Water Works Digest</u> 9, no. 1, 1967 (Hamilton, Ontario: Stanton Pipes [Canada] Ltd.).
4. The change in rate occurs within the average residential consumption.
5. Partly metered.

II

Systematic variations in residential water demand: Empirical evidence and possibilities for planning

FACTORS AFFECTING RESIDENTIAL WATER USE

Environmental Approach

Both the sources of supply of municipal water and its demand are distributed unevenly over time and space. The availability, quality, and cost of a piped water supply depends on the spatial juxtaposition of source, plant, and customers. In order to obtain, develop, and deliver a supply of piped water of the accepted standards of purity, sufficiency, and dependability one has to overcome the friction of distance at some specified cost. The temporal variations in distribution reinforce the spatial variations because the supply has to be evened out to meet the demand. For example, the highest demand on a river may occur at a time when annual flow is at its lowest.

The level of municipal and residential water use and its fluctuations over time and space have been the subject of some discussion in the literature. Taken as a whole these studies constitute the state of the art of predicting and explaining the variations in the level of residential water use. In the discussion of the factors affecting the level of residential water use, emphasis will be put on the value of these studies for theoretical and applied work and particularly on two characteristics: (i) comparability of the reported results in terms of content and methodology and (ii) the validity of the hypothesized relationships.

The nature of the research problem and the general lack of data are in large part responsible for the variety of approaches and the contradictory or inconclusive results. Linaweaver, Geyer, and Wolff put it as follows:

Many factors influence water use in residential areas in varying degrees according to the type of use, the type of service provided in the area, and the period of demand in question. With so many variables, identification of the major factors which affect use depends to a great extent on the design of the data collection program.[1]

It is generally agreed that the level of residential water use is a response to the environmental conditions (i.e., the physical conditions and the economic-managerial decisions varying over time. and space affect the pattern and level of water use). The environmental conditions may be reduced to a number of factors and preferably the effect of a specific factor may be isolated, ceteris paribus. This approach offers several advantages:

1. It is suited to quantitative analysis so that under certain conditions conclusions correct within specified confidence limits may be stated.
2. It explores some of the interrelationships between residential water use and selected environmental factors (some of which are physical, others social and economic; some are subject to management and others are not).
3. Statements of wide applicability may be made about the effects of variables that are not unique to points in time and space.
4. The tentative conclusions not only explore the pattern of man's behaviour in using a basic natural resource but also have implications for improving the quality of life by better planning.

Before discussing the empirical findings of previous studies in this field, it is useful to list the factors that may affect the

[1]F. P. Linaweaver, Jr., J. C. Geyer, and J. B. Wolff, A Study of Residential Water Use (Washington DC: Department of Housing and Urban Development, FHA, February 1967), p. 28.

33

level of residential water per dwelling unit.

Factor	Hypothesized Consequences	Subject to Control by Water Management	Others (e.g., planners)
A. Physical Factors			
Temperature during growing season (mean and distribution)		No	No
Amount and frequency of precipitation during growing season	Affects amount of water required for lawn watering	No	Yes
Water deficiency		No	Yes
Length of growing season		No	No
Water-retaining characteristics of the soil	Affects rate and frequency of watering but not the amount required	No	Yes
B. Family Income: Ability to use water			
Level of disposable income in household	Affects the number and types of water-complimentary activities	No	No
Level of disposable income per capita		No	No
Ownership, frequency of use and efficiency of water-using equipment (e.g., sprinklers, baths, toilets, washing machines, dishwashers, garbolators, swimming pools, and fish ponds)	Affects both the indoor and outdoor uses of water particularly during summer	No	No
Size of lot	Affects amount of water required for watering in summer	No	Yes
Area in lawn or shrubs		Yes (exhortation)	Yes

Sales value of residence	Serves as a measure of income and of water-using equipment (e.g., lawn, bathrooms)	No	No
Assessed sales value of residence		No	No
Connection to public sewer	If drainage is not good, less water is used for lawn watering	No	Yes
Type of housing (single unit, townhouses, high-rise)	Townhouses, high-rise apartments are not metered individually; less water is required for lawn watering.	No	Yes
Density of housing		No	Yes

C. Other Household Characteristics

Number of persons in residence	Requirements of water for indoor use are increased	No	No
Number of working adults in residence	Indoor requirements are affected inversely	No	No
Number of children in residence	Probably indoor and outdoor uses are increased	No	No
Number of days the family is not in residence	Requirements are affected inversely	No	No

D. Type of Service: Accessibility and Quality of Supply

Number of taps: single or multiple	Residences with multiple taps use more water	Yes	Yes
Pressure of water supply	Loss from faulty equipment varies directly	Yes	Yes
Quality of water (e.g., colour, hardness, odour, clearness, taste, temperature	Some indoor uses may decline when quality is not high	Yes	Yes

E. Management

Metering	Reduces use due to more careful use and repair of equipment	Yes	Yes
Size of minimum bill	Water use varies directly	Yes	–

Amount of water allowed with minimum bill	Water use varies directly	Yes	–
Frequency of billing	If bill is small, the water use varies directly; if bill is large the water use varies inversely	Yes	–
Commodity charge (price)	Water use varies inversely	Yes	–
Water rates for frontage, size of lot	No effect	–	Yes
Summer charges	Summer water use varies inversely	Yes	–
Effluent charge (as a commodity charge)	Same effect as price	Yes	Yes
Effluent charge (as a flat rate)	Same effect as non-metering	Yes	Yes
Exhortation and restrictions	Water use particularly at peak periods is reduced; usually supplied as a temporary policy	Yes	Yes
F. Cultural			
Religion	Possibly affects indoor uses	–	–
Habits (e.g., being accustomed to a green lawn)	Possibly affects all uses, especially lawn watering	–	–

Ideally the municipal water supply management would be in a position to select particular variables and their known effect in order to plan effective action to meet stated objectives. For example, some of the variables contribute to better forecasting of water demands[2] (income levels and distribution, type of

[2] Particularly relevant in view of the unofficial schemes to 'export' water from Canada to the United States; see W. R. D. Sewell, "Pipedream or Practical Possibility?", F. E. Moss, "Toward a North American Water Policy," and "A Monstrous Concept — A Diabolical Thesis," in Water: Process and Method in Canadian Geography, eds. J. G. Nelson and M. J. Chambers (Toronto: Methuen, 1969).

housing, size of household, climatic conditions, habits of the consumers). Other factors may be relevant to efforts aimed at reducing the use of water in an emergency (e.g., types of appeal to the public, exhortation, and legal prohibition). Other variables (e.g., commodity charges) may indicate the value of water to the individual consumer and therefore contribute to a more efficient use of the community resources allocated or committed to water supply.

Some Findings of Previous Studies

The discussion of findings that follows is organized by factors or variables rather than by individual studies. The objective is to gain a better understanding of the significance and importance of the environmental factors affecting residential water use. Given the potentially large number of variables affecting the level of water use by an individual consumer or a group of consumers, it is more useful to examine the effect of particular variables rather than to evaluate the extent to which a group of variables taken together explain the variability of water use in a specific study.

The previous attempts to model residential/municipal water use discussed below indicate a wide variety of approaches and of the type of sample data collected. There are also some discrepancies in results. Whatever other criticisms can be made of these studies, they all go beyond the widely used approach of estimating 'requirements' on the basis of professional judgement and past experience; the studies selected all attempt to model water use in terms of explanatory variables.

Climatic Conditions

One of the recent and best known studies of residential water use was carried out at the Johns Hopkins University by Linaweaver, Geyer, and Wolff.[3] In order to obtain more accurate estimates of residential water use, the Federal Housing Agency's Office of Technical Standards sponsored a comprehensive study which included data from subdivisions representing all important climatic regions in the United States. Master-meters were used to record the flow into 41 homogeneous residential areas having 44 to 410 dwelling units. The study areas were 'selected' with respect to climate and economic level with all other factors taken at random.

[3] A Study of Residential Water Use.

37

The study areas included 28 metered residential areas, 8 flat-rate residential areas, and 5 apartment areas. The study areas are further differentiated into those connected to public sewers and those using septic tanks for sewage disposal and into Western and Eastern United States by their location relative to the 100th meridian.

The study separates residential use ($\bar{Q}$) into domestic ($\bar{Q}d$) and lawn watering ($\bar{Q}s$) so that $\bar{Q} = \bar{Q}s + \bar{Q}d$. Domestic use is defined as the total use during the non-sprinkling season. [4]

Sprinkling use ($\bar{Q}s$) is computed as follows:[5]

$$\bar{Q}s \underset{<}{=} 0.6 \; c \; Ls \; (\bar{E} \text{ pot} - \bar{P} \text{ eff}) \tag{1}$$

when precipitation occurs, and

$$\bar{Q}s = c \; \bar{Ls} \; (\bar{E} \text{ pot}) \tag{2}$$

during long periods without rain, where

$\bar{Q}s$ is the average sprinkling use in gallons/day;
$\bar{Ls}$ is the average area of lawn in acres per dwelling unit;
$\bar{E}$ pot is the estimated average potential evapo-transpiration for the
   period of demand in inches of water per day;
$\bar{P}$ eff is the amount of natural precipitation effective in satisfying evapo-
   transpiration for the period and thereby reducing the requirements for
   lawn-sprinkling in inches per day.[6]
c is the constant 2.72×10^4 (to reduce acre-inches to gallons)
0.6 is a coefficient to adjust for the difference between actual evapo-
   transpiration from lawns and potential evapo-transpiration.

The coefficient 0.6 is the one that is most interesting. It is derived from Table 7 where it may be seen that the actual sprinkling amounts to about 60 per cent of the estimated potential requirements for all three time periods (annual, summer, and

---

[4] Ibid., pp. 17-18. This procedure is justified for practical purposes but it makes it impossible to test the hypothesis that "climate...has little effect on domestic use" (p. 33), since the measurements for domestic water use are all for the cool or non-sprinkling period. During the summer (i.e., the lawn-sprinkling period) there may be differences in the domestic water use in addition to the differences in the water use for lawn-watering.

[5] Ibid., pp. 58-59 and discussion in chapters 4 and 5.

[6] An approximate measure of effective precipitation is obtained by assuming "a linear variation of 100% effectiveness for the first inch of rain in a month to zero effectiveness for all rain over 6 inches in a month. Such an approach assumes a maximum effective precipitation of 3.5 inches/month." R.K. Linsley and J.B. Franzini, Water Resources Engineering (New York: McGraw-Hill, 1964), p. 383.

TABLE 7. COMPARISON OF ACTUAL LAWN SPRINKLING AND POTENTIAL LAWN SPRINKLING REQUIREMENTS, OCTOBER 1963-SEPTEMBER 1965

Type of Study Area	Annual	Summer (Inches of Water)	Maximum Day
Metered public water and public sewers			
West (10 areas)			
Potential Evapotranspiration	29.8	11.5	0.25
Potential Lawn Sprinkling Requirement	22.5	11.5	0.25
Lawn Sprinkling	14.0	7.4	0.15
East (13 areas)			
Potential Evapotranspiration	30.3	15.8	0.29
Potential Lawn Sprinkling Requirement	15.0	9.5	0.29
Lawn Sprinkling	7.0	4.7	0.14
Metered public water and septic tanks			
East (5 areas)			
Potential Evapotranspiration	27.8	15.3	0.29
Potential Lawn Sprinkling Requirement	12.4	8.1	0.29
Lawn Sprinkling	1.1	0.79	0.03
Flat-rate public water and public sewers			
West (8 areas)			
Potential Evapotranspiration	25.4	14.7	0.29
Potential Lawn Sprinkling Requirement	14.8	10.3	0.29
Lawn Sprinkling	39.4	24.5	0.51

SOURCE: F. P. Linaweaver, Jr., J. C. Geyer, and J. B. Wolff, A Study of Residential Water Use (Washington, DC: Department of Housing and Urban Development, FHA, 1967), p. 47.

maximum day) in the West and 50 per cent in the East. Hence the coefficient 0.6. It is difficult to evaluate the significance of this coefficient and the effect of climatic conditions on water use by households.

Table 7 has several other interesting features. The average sprinkling use is much lower in the East than in the West; this reflects the generally more humid conditions in the East. The average sprinkling use is much higher in the flat-rate areas (all of which are in the West). Areas with septic tanks are reported to have a much lower rate of lawn sprinkling and the increase in water use when such an area changes to public sewers should be kept in mind by municipal water supply management.

On the maximum day (i.e., during a dry spell) the potential lawn sprinkling and the amount of lawn sprinkling show little difference between the study areas in the East and West (metered and public sewers). This means that the waterworks have to cope

39

with a comparable expected maximum demand in the East as in the West but in the East the waterworks have unused capacity more frequently. In eastern Canada, where conditions may be expected to be comparable to those in the eastern United States, the maximum day demands impose costs on the water user that are required only during the relatively rare dry summer spells.

The data obtained by the Johns Hopkins investigators were used in a regression analysis by Howe and Linaweaver;[7] the effect of climate does not emerge very clearly in this study and throws some doubt on the validity of the formula (1) given above for planning purposes. The results are summarized in Table 8. For the average summer day the chosen 'explanatory' variable is $w_s - 0.6r_s$ i.e., the summer potential evapotranspiration in inches computed by the Thornthwaite method less the summer precipitation in inches; 0.6 of r_s was taken as the effective precipitation. For the maximum day rate of lawn watering the relevant 'explanatory' variable is w_{max} or the maximum day potential evapotranspiration in inches. The multiplicative (parabolic) form of the equation is used so that the regression coefficients are exponents of the variables in natural numbers and also measure elasticity.

The coefficient of $w_s - 0.6r_s$ is significantly different from zero and has the expected sign for the metered areas in the East and the flat-rate areas in the West. On the other hand the coefficient for w_{max} is significantly different from zero in the West (metered and flat-rate) but has the wrong sign in the East (metered areas).

It seems that for the east of the United States (and, by analogy in eastern Canada) the amount of summer rainfall and the frequency of rainfall are more relevant than the maximum day potential evapotranspiration. Although the authors of the study point out that the range of variation of the maximum day potential evapotranspiration is small, the evidence presented for the effect of climate is not conclusive.

Haver and Winter[8] used least squares analysis to estimate municipal water use per capita per day for 13 cities in Ontario

[7]C. W. Howe and F. P. Linaweaver, Jr., "The Impact of Price on Residential Water Demand and Its Relation to System Design and Price Structure," Water Resources Research 3, no. 1 (1967):13-32.

[8]C. B. Haver and J. R. Winter, Future Water Supply in London: An Economic Appraisal (London, Ontario: Public Utilities Commission, January 1963), and Supplementary Comments Re Economic Analysis of Future Water Supply of London (London, Ontario: Public Utilities Commission, May 1963).

TABLE 8. EFFECT OF CLIMATE ON
WATER DEMAND FOR SPRINKLING

| Study Areas | Partial Regression Coefficients and Standard Error | |
	Average Summer Day $w_s - 0.6r_s$	Maximum Day w_{max}
21 areas M & PS	2.07 (0.451)	-0.651 (0.760)
10 areas M & PS (West)	1.47 (0.921)	2.06 (0.826)
11 areas M & PS (East)	2.93 (0.429)	-10.40 (5.010)
8 areas flate rate & PS (West)	3.27 (1.450)	8.84 (2.710)

NOTES: M and PS stand for metered and public sewers.
SOURCE: C. W. Howe and F. P. Linaweaver, Jr., "The Impact of Price on Residential Water Demand and Its Relation to System Design and Price Structure, "Water Resources Research 3, no. 1 (1967):25-6.

including Kingston, Chatham, Waterloo, London, Richmond Hill, and Peterborough. One of the independent variables in the estimating equation was the number of days in June, July, and August with a rainfall of 0.01 inches or more. They report that this variable "does not improve the relationship significantly" but no further details are given.

In an earlier study based on data collected by the American Water Works Association, Fourt[9] found that the residential water use for 1955 in 23 small cities and 21 large cities was related to the number of days of rainfall in June, July, and August 1955 and other variables. Size of city is allowed for by Fourt; he points out that "if aggregates or averages are to be compared, cities with a large proportion of dwelling units in apartment buildings should be treated separately or with caution."[10]

The variables selected by Fourt (Tables 9 and 10) are

X_1,   residential water use in 1000 cubic feet/year/person;
X_2,   price in dollars/1000 cubic feet at 1000 cubic feet per month;

[9] L. Fourt, "Forecasting the Residential Demand for Water," Seminar Paper, Agricultural Economics, University of Chicago, February 1958. Mimeographed.
[10] Ibid., p. 6.

41

TABLE 9. SIMPLE COEFFICIENT OF DETERMINATION (r^2) FOR RESIDENTIAL WATER USE WITH RESPECT TO FOUR SELECTED VARIABLES, 1955

	X_2	X_3	X_4	X_5
23 small cities	(-) 0.265	(-) 0.483	(-) 0.065	(-) 0.019
21 large cities	(-) 0.141	(-) 0.311	(-) 0.373	0.098
44 cities combined	(-) 0.201	(-) 0.355	(-) 0.185	0.006

SOURCE: L. Fourt, "Forecasting the Residential Demand for Water," Seminar Paper, Agricultural Economics, University of Chicago, February 1958, p. 8. Mimeographed.

TABLE 10. REGRESSION COEFFICIENTS FOR RESIDENTIAL WATER USE WITH RESPECT TO FOUR SELECTED VARIABLES, 1955

	Intercept	X_2	X_3	X_4	R^2
23 small cities	5.986	-0.325 (.131)	-0.053 (.015)	-0.203 (.244)	.619
21 large cities	5.329	-0.414 (.073)	-0.026 (.007)	-0.339 (.054)	.839
44 cities combined	5.312	-0.386 (.074)	-0.037 (.007)	-0.305 (.064)	.683

SOURCE: Fourt, "Forecasting the Residential Demand for Water," p. 8.

X_3, number of days of rainfall in June, July, and August 1955;
X_4, average number of persons per meter;
X_5, the total population served.

Tables 9 and 10 show that the residential water demand per head varies inversely with price, number of rainy days in summer and, in large cities, the proportion of people living in apartment blocks (X_4).

The evidence in the four studies discussed is not conclusive. Over small areas (e.g., regions such as the one in Haver and Winter's study) the effect of climatic differences on water use may be considered negligible; in any case the range of variance of the observations on climatic differences are likely to be small and this results in inefficient estimates of the regression coefficient.

On a continental scale (e.g., Fourt's study) the range and variance of observations on climatic differences are likely to be greater and the effect of climatic conditions on water use is more evident (Table 10).

Size of Lawn

Climatic conditions affect residential water use mainly during the summer when water is required for lawn sprinkling. Generally speaking the relationship may be expressed as follows:

Water used on lawn = 'water deficiency' . size of lawn   (3)

It was noted above that Linaweaver, Geyer, and Wolff applied the size of lawn into the computation formulae (1) and (2).

This relationship makes sense in a 'deterministic' way but it is difficult to demonstrate empirically because of the size of lawn, the value of the residence, and the level of income are all highly correlated and are surrogates of the same variable. In the follow-up study by Howe and Linaweaver, size of lawn was included among the variables affecting the demand for lawn watering (Table 11).

Only two equations out of eight — the flat-rate areas — have meaningful and significant partial regression coefficients for this variable. In the other equations (Table 11) the partial regression coefficient is either not significantly different from zero or has the wrong sign. It may be significant that for climatic conditions the partial regression coefficient was significant and

TABLE 11. EFFECT OF SIZE OF LAWN ON WATER DEMAND
FOR SPRINKLING (Multiplicative Form)

| Study Areas | Partial Regression Coefficients and Standard Error of $\hat{b}$ | |
	Average Summer Day	Maximum Day
21 areas M & PS	-0.469 (0.252)	0.201 (0.221)
10 areas M & PS (West)	-0.092 (0.304)	-0.076 (0.256)
11 areas M & PS (East)	-0.793 (0.217)	0.118 (0.307)
8 areas flat rate & PS (West)	1.70 (0.802)	0.943 (0.355)

NOTES: M and PS stand for metered and public sewers.

SOURCE: Howe and Linaweaver, "The Impact of Price on Residential Water Demand," pp. 25-26.

had the expected sign only for the flat-rate areas (Table 8).
This suggests that when the marginal price of water is zero the
response to climatic conditions (i.e., the number and duration
of dry periods) is greater than when the marginal price of water
is positive. In flat-rate areas water use would be proportional
to the size of lawn as assumed in (3) above.

This is particularly relevant since the lawn-watering demand
is a peak demand that imposes costs to meet such demand for
comparatively short periods of time. In an early study, Wolff
and Loos found important differences in peakedness of water use
among neighbourhoods in Baltimore (Table 12).

As already pointed out, such evidence of differences between
neighbourhoods may be related to income differences rather than
to size of lawn per se. The frequency of watering is probably
more closely related to the income levels of the neighbourhood
than to size of lawn.[11] The relationship could then be expressed:

Water demand = f (frequency of sprinkling) = f (income level).   (4)

Such an expression leaves out water deficiency and size of lot (3).
Water deficiency differences emerge on a continental scale as in
Fourt's study above; the size of lot would also be omitted because
most of the variation in water use is also being 'explained' by the
differences in income since within income groups the variation
in size of lot is not likely to be great.

The peak hour rate of use is not subject to policy manipulation
given the present pricing and metering systems because the
householders who water the lawn at the peak hour pay the same
as the householder who waters during off-peak periods. In many
cities waterworks are geared to meeting maximum day demands
rather than peak-hour demands. However, the two measures of
water use are related; Linaweaver, Geyer, and Wolff report the
following relation:[12]

$$\bar{Q}_{(pkhr)} = 334.0 + 2.02\,\bar{Q}_{(mxdy)} \qquad (5)$$

where Q is measured in units of gallons per day per dwelling unit.

___

[11] It is true, generally speaking, that lawns in higher income areas are greener.

[12] A Study of Residential Water Use, p. 45 and figure 11 (p. 46).

TABLE 12. PEAK DEMAND IN RESIDENTIAL NEIGHBOURHOODS

Lot Size (sq. ft.)	Neighbour- hood Type	Dwell- ings	No. of People	(A) Water Use Average Daily (By Record)	(B) (Gallons/Capita/Day) Peak Daily (From Test)	(C) Peak Hourly	(C:B) Per Cent	(C:A)
5,000	older	44	153	60	70	310	440	520
7,200	--	41	143	84	304	1190	392	1420
17,000	new	40	138	94	156	1210	780	1290
26,000	new	16	57	95	461	1870	405	1970

SOURCE: J. B. Wolff and J. F. Loos, "Analysis of Peak Water Demands," Public Works 87, no. 9 (1956):113.

Income and Economic Level

Income or economic level of household is the factor most widely accepted as a determinant of residential water use. There is a problem of defining 'income'; income flow figures for individuals are difficult to obtain and attention has been given to the possession of water-using equipment or other measures of real estate. The use of water in the home requires water-using appliances such as baths, toilets, washing machines, and large lots and the possession of these varies with income.

After reviewing evidence from two Indian cities (Kalyani and New Delhi) Lee concludes that "the demand for [residential] water is conditioned by the ability to use water as well as the accessibility of supply."[13] Lee found that the level of water use is related to income factors such as the number of rooms, the number of taps, the number of servants, and possession of a refrigerator.

For indoor or domestic uses in the United States, Linaweaver, Geyer, and Wolff[14] give the following relation:

$$Q_d = 157.0 + 3.46 \ V \tag{6}$$

Q_d is the average domestic water use in g/d/du
V is the average market value of residence in $1000 per dwelling unit.
r is 0.76 (r^2 = 0.58) and

the standard error from the regression is 33 gallons/day/dwelling unit.

[13] T. R. Lee, Residential Water Demand and Economic Development, University of Toronto Department of Geography Research Publications No. 2. (Toronto: University of Toronto Press, 1967), p. 91; also pp. 76-79, 89-91.

[14] A Study of Residential Water Use, p. 29

45

The average value of residences varied from \$9,000 to \$53,000 and the observations are group averages; both conditions tend to reduce the standard error from the regression.

Howe and Linaweaver, using the same data but including more variables, report elasticities of about 0.3 for domestic water use. The range is greater for summer sprinkling demand: 0.68 in the West and 1.45 in the East of the United States. The partial regression coefficients and their standard errors are given in Table 13.

Income level as measured by the value of residence affects indoor water use by households; the exception is the equation for areas not connected to public sewers. In the East the effect of income level is significant for average summer sprinkling demand but less significant for maximum day sprinkling demand.[15] The corresponding coefficients for the areas in the Western US are not significantly different from zero.

The flat rate areas have highly significant partial regression coefficients from maximum day sprinkling suggesting that when the marginal price for water is zero the water consumer is more sensitive to 'drought' i.e., he is more ready to protect his lawn. This confirms a point made above: demand for sprinkling water in flat rate areas is more sensitive to climatic conditions.

Headley[16] studied the relationship between water demand and income in the San Francisco-Oakland Metropolitan Area; his study is particularly interesting because (a) he used income flow as a measure of income level and therefore his study shows that family income is related to water use in the same general way as value of residence; (b) he has two cross-sectional models and a time series model for each municipality.

He considered precipitation, temperature and price of water to be constant over the area[17] and the study focusses on the

---

[15]In a later paper Howe does not report the partial coefficients for the value of residence for the Eastern US, possibly because these partial regression coefficients were not significantly different from zero at the usual level of significance. C.W. Howe, "Municipal Water Demands," in Forecasting the Demand for Water, eds. W.R.D. Sewell and B.T. Bower (Ottawa: Queen's Printer, 1968).

[16]J.C. Headley, "The Relation of Family Income and Use of Water for Residential and Commercial Purposes in the San Francisco—Oakland Metropolitan Area," Land Economics 39, no. 4 (1963):441–449.

[17]The price range was \$3.29 to \$5.20/1000 cu. feet or 50 to 70 cents/1000 gallons; ibid., p. 442, footnote 2.

TABLE 13. EFFECT OF VALUE OF RESIDENCE ON RESIDENTIAL
WATER USE (Multiplicative Form)

	Partial Regression Coefficient (b)	Standard Error of b
Domestic Demand		
21 areas M & PS	0.352	0.071
13 areas Flat Rate & PS	0.378	0.066
5 areas M & Septic Tank	-0.109	0.141
Summer Sprinkling Demand		
21 areas M & PS	1.070	0.299
10 areas M & PS (West)	0.685	0.396
11 areas M & PS (East)	1.450	0.306
8 areas Flat Rate & PS (West)	0.447	0.248
Maximum Day Sprinkling Demand		
21 areas M & PS	0.405	0.266
10 areas M & PS (West)	0.438	0.322
11 areas M & PS (East)	0.931	0.443
8 areas Flat Rate & PS (West)	0.395	0.099

NOTES: M and PS stand for metered and public sewers.

SOURCE: Howe and Linaweaver, "The Impact of Price on Residential
Water Demand," pp. 24-26.

income variable. Headley reports the following relationship:

$$X_o = -30.24 + 2.16X_1 \quad \text{for 1950} \quad (r^2 = 0.81)$$
$$(0.30)$$
$$X_o = -18.77 + 1.27X_1 \quad \text{for 1959} \quad (r^2 = 0.80)$$
$$(0.19)$$

where X_o is gallons/capita/day purchased and

X_1 is the median family income per year in $100.

The two estimates of b were found not to differ significantly
i.e., "the cross-sectional income parameter was probably un-
changed over the time period in question." This result is to be
expected because cross-sectional analysis reflects long-run
behaviour, i.e., it is assumed that consumers have adjusted
their water use to their level of income.

The income elasticity of demand for residential water was calculated from the linear equations and from the log-linear equations as follows:[18]

	1950	1959
Linear equations (at the means)	1.49	1.24
Log-linear equations	1.63	1.37

In general an increase in income by 1 per cent will increase water use by about 1.5 per cent. Two further comments need to be made. These relationships make the assumption that the spatial variation in family income is not related to any other potential explanatory variable such as price, number of persons in residence, size of lot, and so on. In other words all the variation in water use that can be explained by variation in income is ascribed to income. This may be part of the reason why Headley's results are significant for part of the western US when Howe and Linaweaver's were not significant for the West (Table 13).

Secondly these elasticities were estimated from cross-sectional (or spatial) data and will not usually apply to time series data. Headley's study provides an illustration on this. Only two of the 11 equations based on time-series data have $\hat{b}$'s that are significantly different from zero at the 10 per cent level of confidence, none at 5 per cent. The average income elasticity is about 0.2. This suggests that the acquisition of water-using appliances is a long-term adjustment; the year to year increases in real income have only a limited effect on year to year increases in residential water use. Many water-using appliances are consumer durables (baths, washers, swimming pools) or even once-in-a-lifetime purchases like a large lot for a residence.

The time-series low income elasticities and the cross sectional high income elasticities can be interpreted as follows: a neighbourhood with a high median income may be expected to include a high percentage of families which possess water-using appliances and large lawns; as the income increases over time, the level of water use increases at a slower rate because many of the durable consumer goods to which water is complementary have already been made.

[18]Ibid., pp. 444-445.

Turnovsky[19] reports planned residential water use to be significantly and positively correlated with a measure of per capita housing space. The measure is:

$$h_i = \frac{\text{Average number of rooms per dwelling unit in town i}}{\text{Median number of occupants per dwelling unit in town i}}$$

This measure reflects real estate availability but the results are not comparable to those reported in other studies except as a strong indication that inter-urban differences in the level of water use by households may be partly attributed to income levels.

Number of Persons in Residence

Given the climatic conditions and the income level, the amount of water used in a household or community can be written as an identity:

$$\text{Water use}_i = n_i \bar{q}_i \tag{7}$$

where q_i is the average per capita water use and n_i is the number of persons in a residence. This identity is implied in several studies where the dependent variable is the per capita water use. It is also a basic assumption in the forecasting procedures currently in use which will be discussed in the next section.

This assumed linear relationship between the number of persons and water use--like others examined above--does not emerge as a universal finding in previous work. Howe and Linaweaver[20] found a significant relationship between water use within the house and size of household for flat-rate and apartment areas and septic tank areas but not for metered areas with public sewers. This last category of consumers is the most important numerically since most new subdivisions are metered. A possible explanation for this unexpected finding is that metered areas had a low variance for n_i (number of persons per dwelling unit) with a resultant inefficiency in the estimator $\hat{b}$. The number of persons/dwelling unit ranged from 4.7 to 1.8 in the flat-rate areas and from 2.7 to 4.9 in the metered areas.

---

[19]S. J. Turnovsky, "The Demand for Water: Some Empirical Evidence on Consumers' Response to a Commodity Uncertain in Supply," Water Resources Research 5, no. 2 (1969):350-361.

[20]"The Impact of Price on Residential Water Demand," pp. 18 and 24.

Metering

The effect of metering on the level of residential water use has been the subject of some controversy. Some of the disagreement is due to the confusion of three separate questions:

1. Is residential water use lower in those municipalities where the water is metered? Which residential uses are reduced most?
2. Does water use decrease permanently following metering?
3. What is the proper criterion to use in deciding whether metering ought to be introduced?

The first two questions will be discussed at this stage; the third question will be taken up in Chapter IV as part of the review of a recent study on whether to introduce metering in the City of Toronto.

1. Effects of metering

Metering per se need have no effect on water use.[21] It is the variable price attached to metering that results in a response by the consumer. The comparison is not between metered water use generally and unmetered use but among variable marginal prices ranging from zero (for unmetered water use) to one dollar or more per 1000 gallons. For example, the evidence presented in Table 7 is incomplete. The average variable price for water of the metered areas is about 40 cents and ranges from 14 to 102 cents/1000 gallons.[22] In other words the comparison is between water use in areas where water is 'cheap to expensive' and water use in areas where the marginal price of water is zero. The comparison between cheap metered water and unmetered water may be entirely different.

Metering has most effect on the less essential water uses such as lawn sprinkling (see Table 14) domestic or indoor uses are hardly affected. Sprinkling and related uses affect the maximum day and peak hour use to a much greater extent than domestic use and the peak uses are most relevant to design and planning. Therefore metering may reduce the need for storage capacity installed to meet peak demands.

---

[21] This statement will be tested against a sample of consumers from the Toronto area in chapter III.

[22] Data in Howe and Linaweaver, "The Impact of Price on Residential Water Demand," p. 18, table 2.

TABLE 14. METERED AND UNMETERED
RESIDENTIAL WATER USE, WEST UNITED STATES

	Metered & Public Sewers (10 Areas)	Flat Rate & Public Sewers (8 Areas)
Domestic Use		
Mean annual use	247	236
Mean of maximum day uses	454	431
Mean of peak-hour uses	1214	1016
Sprinkling Use		
Mean annual use	186	420
Mean of maximum day uses	707	2083
Mean of peak-hour uses	2076	4812
Residential Use (Domestic and Sprinkling)		
Mean annual use	458	692
Mean of maximum day uses	979	2354
Mean of peak-hour uses	2481	5170

SOURCE: Linaweaver, Geyer, and Wolff, A Study of Residential
Water Use, tables 2, 3, 4.

The rate of use of water for lawn sprinkling in unmetered
areas is almost three times the rate of water use in metered
areas on the maximum day. That lawn sprinkling in flat-rate
areas is inefficient is shown by the fact that the rates of actual
lawn-sprinkling are over twice the rate of sprinkling requirements
as opposed to 0.6 in metered areas (Table 7). It is significant
that on an average summer day the coefficient of lawn watering
with respect to water deficiency in the West of the United States
in non-metered areas is double that in metered areas; on the
maximum day it is about four times as much (Table 8).

The waterworks industry is aware of the possibility of
reducing water demand by means of metering new subdivisions.
For example, the 1962 edition of the Canadian Municipal Utilities:
Waterworks Manual and Directory[23] claims that 100 per cent
metering reduces water consumption by 40 to 60 per cent.

---

[23]Canadian Municipal Utilities: Waterworks Manual and Directory, 1962 (Toronto: Monetary
Times Publications Ltd., 1962), p. 57.

2. Does metering reduce water use permanently?

Pitblado throws some light on the second question above, viz., does the introduction of metering decrease water use permanently or is the decrease temporary? He summarizes the evidence from St. Catharines, Ontario:

...the average annual consumption of water per residence was reduced by 11% when meters were installed. However, the increase in water bills after metering was not sufficient to cause alarm. As persons became used to having meters in their homes, they relax their drive to conserve water. As a result the average consumption of water in 1965 was higher than it was in 1963, just after metering and not significantly different from consumption per household prior to metering.[24]

It is instructive to note that in St. Catharine's the minimum quarterly bill was $4.00 for 8000 gallons (or 50 cents per 1000 gallons) and the marginal price was 26 cents/1000 gallons. This minimum bill was advocated because it allows small water users a low-cost allowance of water. This minimum allowance not only penalizes the small user of water but it reduces the incentive to conserve water following metering as will be shown below.

Hanke and Flack[25] quote evidence from Boulder, Colorado in support of their contention that the introduction of metering reduces the level of water use (Table 15). The annual water use is reduced by 34 per cent and the summer use is reduced by 37 per cent if one takes a mid-point between the water use in a dry year and a wet year. The lag between the introduction of metering and the time of the study was short and water use may increase after a few years.

Increases over time following the introduction of metering may be due to factors other than the relaxation of consumers. Hanke and Flack point out that the increase observed a few years following the introduction of metering may be a shift to the right of the demand function in response to increases in income for example. However, the work of Headley discussed above suggests that such a shift in demand is likely to be very small; an increase of 1 per cent per annum in income resulting in water use increase of 0.2 per cent and if his results are accepted an alternative explanation may be presented.

[24] J. R. Pitblado, "The Effects of Metering on the Domestic Consumption of Water — The City of St. Catharine's," B.A. Thesis, Department of Geography, University of Toronto, 1967, p. 46.

[25] S. T. Hanke and J. E. Flack, "Effects of Metering on Urban Water," Journal AWWA 60 no. 12 (1968):1359-1365.

TABLE 15. EFFECTS OF METERING, BOULDER, COLORADO

Year	Per Cent Metered	Annual Use	Winter Use (Gallons Per Capita Daily)	Summer Use
1960	5	243	154	365
1965 (wet year)	100	149	107	206
1964 (dry year)	100	172	111	257

SOURCE: S. T. Hanke and J. E. Flack, "Effects of Metering on Urban Water," Journal AWWA 60, no. 12 (1968):1364.

The increases over time following the introduction of metering are due to low marginal prices and high minimum bills. For example, if a family has an allowance of 8000 gallons a quarter with a minimum bill of $4.00 and requires about 2000 gallons more at 30 cents/1000 gallons, the variable bill is only 60 cents. This is too small a change in the water bill to act as an incentive to reduce water use.

In general, unmetered consumers have no incentive to use water efficiently or to repair indoor water-using fixtures. Total residential water use is about 30 to 50 per cent higher in flat rate areas, with most of the extra demand occurring during seasonal peaks for lawn watering.

Pricing
The demand curve for most commodities shows the quantity bought (Q) to be a declining function of price (P). This relationship is not widely accepted in the water utility industry because, it is argued, water is a small item in the family budget and there are hardly any substitutes for water. Under present pricing systems a large proportion of the water bill is fixed irrespective of the amount of water used and the marginal price of water is often low. Some of the studies of residential water use in recent years have assumed that "the demand for water shows a systematic variation conditioned by a number of factors in the living environment — a relationship not dependent upon the price of water."[26]

[26] Lee, Residential Water Demand and Economic Development, p. 4. Similar assumptions are made in the study by Linaweaver, Geyer, and Wolff and in a forthcoming study by Gilbert F. White, D.J. Bradley, and Anne U. White, Drawers of Water: Domestic Water Use in East Africa (in press).

The effect of price upon water use is of basic importance to residential water management. Price setting is one of the few instrument variables at the disposal of the management and prices may be used to allocate resources efficiently in publicly controlled monopolies such as municipal waterworks. There is one problem of definition of 'price' which should be discussed at the outset.

Residential water is often priced on a declining block rate schedule so that the price per 1000 gallons declines as the amount used increases (Figure 3). Municipality B in Figure 3 has a fixed rate or minimum bill for the first few units and then one common rate, so that all consumers, irrespective of how much they use beyond the minimum bill, are facing the same price.

Municipality A has a fixed rate for the first few units and then two blocks with a reduction in price around the 120 gallon/day/dwelling unit level (say 12,000 gallons/quarter). This means that 'small' consumers are paying a marginal price of P_1 and 'large' consumers are paying a marginal price of P_2, when $P_1 > P_2$.

When data for individual households from municipalities such as A are plotted, there is no other way for the demand curve to be drawn except as a declining function of price. The observations logically would describe a situation where

$$P_i = f(Q_i)$$

rather than confirm a hypothetical relationship such as

$$Q_i = f(P_i)$$

This pricing system for residential water creates some problems because if municipalities of Type A are included in the sample of observations, the resulting demand curves may simply reflect the fact that low prices are charged when water use is high and vice-versa.

When the observations consist of averages for municipalities the problem still remains. Assuming for purposes of example that the water use in one municipality falls within the lower price block P_2 (i.e., $Q > Q_1$) and the average water use in another municipality falls within price block P_1 (i.e., $Q_0 < Q < Q_1$), then a line such as DD cannot be drawn except as a declining function of P even if Q is not related to P. In other words the null hypothesis that $b = 0$ cannot be rejected in the case of Type A municipalities. A hypothetical example of two municipalities

54

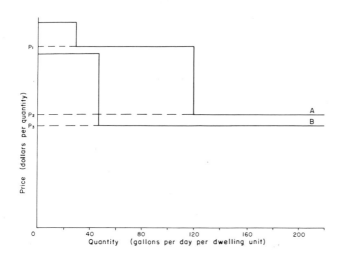

Figure 3. Two Types of Declining Block Rates

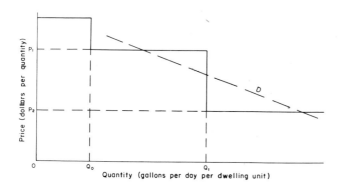

Figure 4. Spurious Demand Curve from
Pricing Schedule A in Figure 3

55

of Type A is given in Figure 5. In municipality A_1 the average water use is low, in municipality A_2 the average water use is high (e.g., it is in a more affluent district). Then A_1 must have a high price associated with low use and vice-versa for A_2, but the relationship is a spurious one if price varies with quantity. [27]

The studies of residential water use discussed below all make use of average figures but none state whether the sample includes municipalities of Type A in Figure 3. It is therefore impossible to determine to what extent, if any, the price-quantity curve is true or spurious. In most municipalities it is reasonable to assume that residential water users fall within one wide price block and therefore most of the observations would reflect the position B in Figure 3.

With that caveat one can point to several studies that report that the amount of residential water used is a declining function of price. Howe and Linaweaver found that the price elasticity of water is least for indoor use (-0.23). It is highest for sprinkling use in the Eastern US (-1.57) and about -0.7 for sprinkling use in the drier Western US. [28] The elasticity for sprinkling use in the Eastern US was revised in a later report by Howe to -0.9. [29] Fourt reported a significant regression coefficient for price (Table 9). At the means the elasticity is -0.4 which is lower than those reported by Howe and Linaweaver.

Haver and Winter [30] hypothesized a constant elasticity model (i.e., linear in the logarithmic form) and report a price elasticity of -0.254; this seems to be low for Ontario but in any case it is based on an average price for all municipal water and has to be interpreted with caution since large-water-using plants are likely to be coincident with relatively cheap water. A similar study for cities in Kansas by Gottlieb[31] also reports price elasticities of

[27] Howe and Linaweaver, "The Impact of Price on Residential Water Demand," pp. 21-23 discuss another aspect of the spurious demand function, viz., that if supply costs are similar from utility to utility, the utilities with high water use per dwelling unit can afford to charge lower prices and vice-versa. The problem is not considered in detail due to lack of data about costs.

[28] Ibid., pp. 24-25.

[29] Howe, "Municipal Water Demands," p. 70.

[30] Future Water Supply of London: An Economic Appriasal, p. 13.

[31] M. Gottlieb, "Urban Demand for Water: A Kansas Case Study," Land Economics 39, no. 2 (1963):204-210.

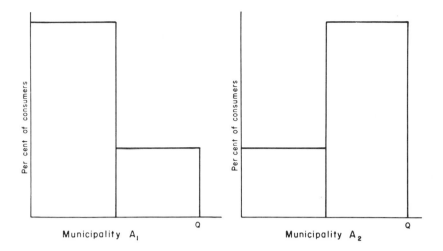

Figure 5. Hypothetical Distributions of Residential Water Consumers in Two Municipalities of Type A in Figure 3

-0.4 and -0.6. Since this study was based on average <u>urban</u> water use and average price (total revenue/total water use) the results cannot be applicable to the marginal pricing of residential water use.

Turnovsky[32] reports that the <u>planned</u> per capita water use by households is dependent on price. He uses average consumption and average prices for a group of towns in Massachusetts. The price elasticities are reported to be from -0.25 to -0.4 at the means which is in general agreement with other studies based on average figures for municipalities (e.g., Fourt, Haver and Winter), but these elasticities are lower than those obtained by Howe and Linaweaver from sample data for residential consumers.

The results discussed above are comparable only in a very general sense. The investigations refer to different measures from different groups of cities, for different time periods, using different functional forms of the equation.

[32]"The Demand for Water," p. 356.

PROJECTING THE DEMAND FOR RESIDENTIAL WATER

Society is constantly striving to plan better. This may be done through discovering by research the relevant variables, relationships, and distinguishing characteristics that will replace or supplement opinion, vague generalization, and rules of thumb. This section will focus on setting out the principal components of information that should form part of better forecasting and planning of residential water use. Information about the projected needs or demands for residential water during peak periods is basic to the planning of water supplies; the maximum day demand plus fire use, or some other measure of the 'requirements' appropriate to the locality and to the facility being built could be taken as the target variable in the discussion.

It would be assumed that in general water is available at a cost and that more water is <u>always</u> available but at a higher cost. In other words the shortages are in the facilities not in the water itself.[33] Failure to match supply with demand results either in some loss of utility (if the supply is not enough) or in over-investment.

There is considerable urgency in improving the planning and projection of residential water demand in view of the vast programme of building water supply facilities that lies ahead. According to Howson:

In the past 10 years as much money has been invested in water supply in the U.S. as in all preceding years. In fact, 3/4 of the total historical national water supply investment measured in dollars has been made since 1943.[34]

In spite of the millions of dollars spent each year for water-works, the municipal water utility spokesmen are not satisfied. The President of the American Water Works Association said in 1960:

We are in the midst of a critical shortage of water facilities. Frankly we are not properly equipped to collect, store, process and transport water in sufficient quantity and quality to meet the needs of our population.[35]

[33] R. J. Faust, Executive Secretary for the American Water Works Association, uses the same argument in a plea to take measures to prevent another large-scale drought; "N. E. Water Crisis," Journal AWWA 58, no. 1 (1966):3.

[34] L. R. Howson, "Revenues, Rates, and Advance Planning," Journal AWWA 52, no. 2 (1960):153-154.

[35] L. W. Grayson, "Water Supply — America's Greatest Challenge," Journal AWWA 52, no. 1 (1960):2.

The reasons for this increase in the need for facilities are:

1. Larger populations living in cities.
2. Increasing per capita demand.
3. As cities increase in size they often have to tap new sources of water so that transportation costs increase; this factor may be partly or wholly offset by some economies of scale.
4. Increasing peak demands so that more storage is required to meet seasonal, day, and hourly peak demands.
5. Increase costs of distribution due to lower population densities of suburban areas.

Review of Present Practice

One may add that the forecasting methods practiced at present frequently overstate the demand and this is recognized by some engineers in the water utility business. For example, Flack argues that this over-estimation of demand is due to making four simplifying assumptions which can be seriously misleading:[36]

1. Technological changes will not occur and in view of thier uncertainty it is safer not to rely on them.
2. Price increases will not occur, or if they do, they will not affect demand; therefore 'requirements' are forecast and then water rates are set to recover the expenditures.
3. Inefficient uses will continue (e.g., over-watering of lawns, losses through mains leakage).
4. Industrial expansion and water use will grow in proportion to population growth.

Technological advances are difficult to forecast and the first assumption is required to make the forecast feasible. The fourth assumption referred to by Flack could be removed if detailed industrial projections are available;[37] metropolitan cities often have detailed land use forecasts which may be used in forecasting water demand/requirements. Unfortunately it is still common to come across official projections of total urban water use which imply that all water use is a function of population; some examples are discussed briefly below.

[36]J.E. Flack, "Meeting Future Water Requirements Through Reallocation," Journal AWWA 59, no. 11 (1967):1340.

[37]See Howe, "Municipal Water Demands," p. 63 ff, for a discussion of one type of projection — economic base studies.

59

The two other assumptions were made less frequently by forecasters of water use in the 1960's. Metering is now widely accepted as a management tool and inefficient uses of water (e.g., leakage from mains) are usually cut when the cost of supply makes such attempts worthwhile.[38]

By changing the assumptions listed above (e.g., deciding to meter residential water use or reduce losses through mains leakage or to prohibit the use of water-using air-conditioning equipment),[39] the water utility management can influence the per capita or per acre rate of water use; therefore policy variables should be included in the projection.

There are two fundamentally different approaches to water use projection. The most commonly used is the requirement approach: the amount of water 'required' is determined by one variable (e.g., population or number of dwelling units or the area of land in residential use) which measures the intensity of demand. The magnitude of the variable is then multiplied by a per unit rate of use. The procedure is flexible enough to allow for increases in the per unit rate of use over time and to take into account the ability of the consumer to purchase water by distinguishing among neighborhoods or user groups.

A basic assumption of the requirements method is that the price of water or any other management tool does not affect the amount of water purchased. It is this basic assumption that differentiates the requirements approach from the second approach which includes price and other management tools as a determinant of water use. For want of a better term this second approach will be called the 'economic demand' approach.

Both approaches need information about (i) the probable magnitude of the determining factors of water use such as population or area of land in residential use and (ii) appropriate coefficients that may be applied to the determining factors. The difference lies in the number of variables taken into account and this, in turn, reflects the theoretical basis from which the projection stems.

---

[38] See Pitblado, "The Effects of Metering...St. Catharine's," and City of Toronto, Department of Public Works, Report on City of Toronto Water Distribution System, (Toronto, April 1968).

[39] L.D. Kempton, "Air Conditioning Brings Water Problems," Public Works 87, no. 9 (1956):132-134.

The forecast of the magnitude of the variable that measures
the intensity of demand does not primarily concern the water use
forecaster; economic base studies and population/land use
forecasts are usually available from planning offices. It is enough
to note that it is by no means a trivial task and any errors in this
phase affect the final result proportionately.[40] Even if policy
variables are taken into account "population growth in combination
with per capita consumption is the most important factor affecting
total future municipal water requirements."[41]

For example, in Figure 6, if the population projection of
350,000 people for London, Ontario in 1985 is accepted, the
'required' water supply is 68 MGD on the maximum day. The
lower population estimate proposed by Haver and Winter makes a
corresponding requirement of 54 MGD; it would delay the building
of phase II of the planned investment and defer to a much later
date the need for phase III.

The main task of the water demand forecaster is to furnish
information about (i) appropriate coefficients of use per person
or per dwelling unit or per acre for different categories of users
(single unit dwellings/apartment dwellings, high/low income);
(ii) changes in the rates of use over time and (iii) peak rates and
their frequency. The present methods of water forecasting are
particularly naive in accepting per unit rates of water use for
large urban areas and in neglecting control of policy factors but
they are improving as the following examples show.

In a survey carried out by Public Works in 1956 the entries
from the engineering consultants and waterworks superintendents
give a cross-section of the approach of prevailing practices.
One reply reads:

Estimating future consumption is a matter of educated guesswork based on
a study of local conditions and past experience. We believe water consumption of individuals in homes will continue to increase at the rate of 1%
per year, and that these individuals will be using, for home purposes, 50
GPCD in 1966 and 55 in 1976. This does not include commercial and industrial uses. [The next reply states] ... estimate that consumption in
gallons/day/person will increase 2% per year for the next 20 years.[42]

[40]See Howe, "Municipal Water Demands," p. 63 ff, for a review and a bibliography on
this phase of the forecast.

[41]US Congress, Senate Select Committee on National Water Resources, Future Water
Requirements for Municipal Use, 86th Cong., 2nd Sess., Comm. Print No. 7 (Washington
DC: GPO, 1960), p. 9.

[42]"Present and Future Estimates of Water Consumption," Public Works 87, no. 12 (1956):
74.

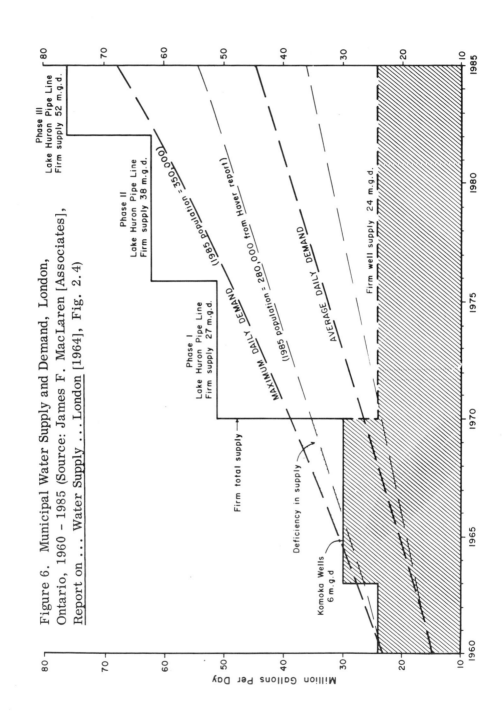

Figure 6. Municipal Water Supply and Demand, London, Ontario, 1960 – 1985 (Source: James F. MacLaren [Associates], Report on ... Water Supply ...London [1964], Fig. 2. 4)

For Ontario, Watt allows 50-60 gallons/capita/day (GPCD) for residential areas, 100 GPCD for areas having some industry and 200 GPCD for areas having heavy-water using industry, each large water plant being allowed for separately. [43]

The Minimum Design Standards acceptable to the Federal Housing Administration as revised in July 1965 recommends that (1) where experience data are available the average demand should be ascertained on the basis of existing systems of a similar nature in the area; (2) in the absence of data the average demand is assumed to be 400 gallons per day per dwelling unit (G/D/DU); the maximum daily demand of 200 per cent of the average is recommended (i.e., 800 G/D/DU) while the maximum hourly demand of 500 per cent of the average demand (i.e., 2000 G/D/DU) is suggested except in areas of extensive lawn irrigation where 2800 G/D/DU are recommended. [44]

The views of the US Senate Select Committee on National Water Resources merit comment both because they reflect an 'official' point of view of management and also because the Committee projections are the best available for Anglo-America so far on a national basis; such exercises are extremely useful in pointing out lacunae in the methodology, apart from the forecasts of water use themselves. The committee exemplifies a common bias among water utility management towards over-investment:

Considerable judgment is, therefore, required in planning for future services, but an important consideration to be kept in mind is that, in developing public water supplies, it is essential that every precaution be taken in the interest of public safety and welfare. For these reasons, it is generally better to overbuild than to underbuild. [45]

The Committee's projections are based on "extrapolation of past trends and upon best judgements as to probable future requirements."[46] The Committee discussed the "effects of efficient management upon municipal water use," and noted that:

there are certain water system management practices that can be undertaken to control water consumption. Leakage surveys, inspection of plumbing, and metering, coupled with consumer education, will control wastage. [47]

[43] A. K. Watt, "Adequacy of Ontario's Water Resources," The Canadian Mining and Metallurgical Bulletin 60, no. 664 (1967):920.

[44] Quoted by Linaweaver, Geyer, and Wolff, A Study of Residential Water Use, pp. 3-4.

[45] Future Water Requirements for Municipal Use, p. 9.

[46] Ibid.

[47] Ibid., p. 13.

Pricing as policy is dismissed the argument being offered that:

...there is really plenty of <u>water</u> for municipal purposes, and the objective of metering should be to prevent waste and to provide a basis for charging the customer for what he uses, not to deliberately restrict his consumption.[48]

The Committee suggested that while

with most products, there is a relationship between price and demand; if the cost increases, the demand decreases. This is also true with water but the limits are not clearly defined. Water plays a unique role in that it is a vital necessity for life and yet is probably the cheapest commodity we buy, cheaper than dirt. Thus, should cost be even tripled, water would still be cheap and long-range domestic demand would probably not be significantly depressed.[49]

The Committee's projections, in other words, are based on the assumption that use should not be limited deliverately and that it is preferable to overinvest in municipal water supply than to underbuild and run short.

A similar projection of water use in Canada describes the methodology as follows:

Referring to American sources, it was assumed that consumption per capita in urban, residential and commercial categories will increase at an approximate rate of 2% per year. Projections of population for each region on the basis of recent trends, adjusted to be consistent with the Gordon Commission forecast were used in conjunction with per capita consumptions to obtain the total residential and commercial consumption up to 1990.[50]

The Ontario Water Resources Commission forecasts water demand in its Report of Water Resources for each county. Extrapolation of past trends and best judgement of future requirements are used in such exercises. For example:

The percentage increase in water consumption from 1936 to 1957 was 110%. Therefore for the next 20 years new treatment facilities should be based on at least double the present pumpage (i.e., double 3.765 MGD on an average day).[51]

---

[48] <u>Ibid.</u>, pp. 15-16.

[49] <u>Ibid.</u>, p. 16.

[50] D. Cass-Beggs, "Water as a Basic Resource," in <u>Water: Process and Method in Canadian Geography</u>, eds. J.G. Nelson and M.J. Chambers (Toronto: Methuen, 1969), pp. 17-18.

[51] Ontario Water Resources Commission, "Report on Water Resources Survey, County of Kent," (Toronto, October 1958), p. 15. Mimeographed.

The forecast for the Windsor area is set out in another report as follows:[52]

Population in 1957	55,000
Increase in population 1937-57	370 per cent
Expected population in 1976 on this basis	258,000
1937 consumption rate	52 GPCD
1957 consumption rate	54 GPCD
Expected 1976 consumption rate	100 GPCD
Water requirements (i.e., 100 x 258,000)	25.8 MGD

At its best the requirements approach can be a very useful exercise. A recent example from Ontario is the forecast for London, Ontario. The method followed by the consulting engineers is more detailed than that followed in other reports and it may be summarized as follows:

1. Project population and area development to target data (P);
2. Compute the rate of water use per capita or per acre for various types of use (residential, commercial, industrial, [b]);
3. Compute the trend of increase in the rate of water use (r);
4. Compute the expected rate of use i.e., rb;
5. Projected water use = Prb;
6. Apply the present or 'best guesses' ratios of
   (i) maximum day to average day use (1.5:1.0)
   (ii) peak hour to average day use (4.0:1.0).[53]

This procedure is an improvement on the other forecasts discussed above because the water requirement is projected for each type of use; however, the coefficients are still fixed. No allowance is made for varying rates of use within the same class (residential, industrial) or for management changes. The previous discussion on the factors affecting residential water use showed that coefficients do vary a lot. Table 16 compares the standards set by the Federal Housing Agency with the more detailed coefficients obtained by the Johns Hopkins investigation on Residential Water Use. The FHA standards are remarkably close to the

---

[52]Ontario Water Resources Commission, "Report on Water Resources Survey, County of Essex," (Toronto, March 1959), p. 14. Mimeographed.

[53]James F. MacLaren (Associates), Report to the Public Utilities Commission of London on Waterworks Developments to 1985 (London, Ontario: Public Utilities Commission, December 1961).

TABLE 16. COMPARISON OF FHA STANDARDS AND FINDINGS OF JOHNS
HOPKINS RESIDENTIAL WATER USE STUDY (Gallons/Dwelling Unit/Day)

| | FHA | Johns Hopkins Residential Water Use Study | | | | |
		National Average	West M & PS	East M & PS	Flat Rate (West)	Metered Water & Septic Tanks
Daily Average	400	389	458	310	692	245
Maximum day Average	800	870	979	786	2354	726
Peak hour Average	2000	2115	2481	1833	5170	1835

NOTES: M and PS stand for metered and public sewers.

SOURCE: F. P. Linaweaver, Jr., J. C. Geyer, and J. B. Wolff, "Summary Report
on the Residential Water Use Project," Journal AWWA 59, no. 3 (1967):269-278.

national average obtained by the Johns Hopkins investigators.
However, the adoption of average standards would over-estimate
the demand (and the investment) in the East and underestimate it
in the West.[54] In addition the Johns Hopkins study provided
coefficients for the income level and the water deficiency and
therefore the projection for individual municipalities can be made
more detailed.

If one relaxes the assumption of fixed required water inputs,
one can introduce management variables such as price, metering,
and restrictions. This should become more important since as
the living standards rise, the use of water becomes less a
matter of necessity and more of increasing satisfaction or
utility.

Some projections which include price as a variable were
discussed above, e.g., Fourt, Haver and Winter, Howe and
Linaweaver. These constitute a definite advance on the requirements approach, and as the costs of water supply increase such
projections should become more appropriate. In addition to

[54] The FHA standards were revised as a result of the study completed by Linaweaver,
Geyer, and Wolff. When no experience data are available "the criteria developed in June
1966 FHA Technical Study Report... may be used." US Federal Housing Administration,
Mimimum Design Standards for Community Water Supply Systems, HUD Handbook, FHA
4517.1 (Washington DC: GPO, May 1968), p. 52.

variable price and metering, there are other policy variables
that could be varied and the water use supply planner should
consider as many of the alternatives as possible (Table 17 below).

RESIDENTIAL WATER DEMAND PROGRAMMING

The projections of water use obtained by the methods discussed
above still omit important considerations which are pertinent to
planning water resources for the cities:

1. No distinction is made between time-series and cross-sectional
projections.
2. Demand as well as supply has a probability distribution so
that the projected demand (supply) can be exceeded (unavailable)
in some years. To reduce the probability of deficiency more
heavy investment is required and this extra cost should optimally
be balanced by the losses from urban drought that would otherwise
occur.

Cross-Sectional Versus Time-Series Coefficients
Cross-sectional coefficients are readily applicable to new subdivisions or towns at the present time; time-series projections
are relevant to decisions about extensions to the system of a
city to provide for growth.

The first type of projection involves the forecast of the
magnitude of each variable on the right hand side of the equation
(e.g., price, number of persons in residence). On the other
hand, projecting the residential water demand at a future time
period requires in addition the forecasting of appropriate
coefficients. For example, the coefficient for the variable Np
(number of persons in residence) may increase as water-using
gadgets become more generally available with rising income.

Ideally one should (a) project the water use for the municipality
as if it were a new subdivision or town by using cross-sectional
coefficients and then (b) adjust the projection by taking into account
time-series that may be available.

Headley's study of water use with respect to income
illustrates the difference between time-series and cross-sectional
coefficients. The time-series elasticites averaged 0.2 while
the cross-sectional elasticities were about 1.6. A suitable

predictive equation would be:

$$\text{Projected water use } t = N \, (b_1 \dot{Y}_{t_o} + b_2 \Delta Y_{t-t_o}) \tag{8}$$

N is the projected number of units (e.g., dwelling units),

Y and ΔY the projected income and the change in income respectively,

b_1, b_2 the cross-sectional coefficient and the time-series coefficient

Deficiency Probabilities

Unless the source of supply is a comparatively large body of water, the supply conditions vary from year to year; peak demands also vary from year to year with the climatic conditions. Both are sources of deficiency in the water supply.

The level of investment and the target level of water supply at peak demand periods should vary in response to two sets of factors (a) the probability distribution of yield from a basic water source and (b) the costs to society of failure to meet peak demands. The level of investment also varies with the scale economies that may be obtained from increasing the size of additions to the system and the interest on investment for excess capacity.[55]

In Figure 7, at a given cost of supply (P_1) a certain quantity of water can be counted upon 98 per cent of the time (Q_1) while a greater quantity of water can be supplied with 95 per cent probability ($Q_2 > Q_1$).[56] The lower diagram (Figure 8), illustrates the same principle with respect to demand: Q_1 may be expected more often than Q_2 if the price is set at P_1. The same quantity of water can be supplied more cheaply if the probability of deficiency is high; the price at which the same quantity is demanded is higher as the probability of that demand occurring becomes lower.

One can work out the expected annual damage due to drought (a) when supply fails to yield enough water, (b) when demand

---

[55] See the discussion by R.F. Scarato, "Time-Capacity Expansion of Urban Water Systems," Water Resources Research 5, no. 5 (1969):929-936.

[56] Other determinants of cost are the size of the reservoir and the ratio of required flow to mean annual flow. The supply side of deficiency is excellently treated in G.O.G. Lof and C.H. Hardison, "Storage Requirements of Water in the U.S.," Water Resources Research 2, no. 3 (1966):323-354 especially table 5 (p. 337) and table 6 (p. 338).

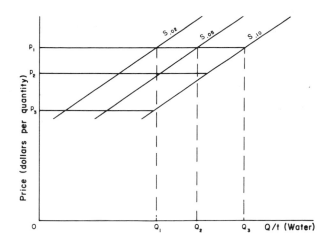

Figure 7. Supply of Municipal Water at
Different Probabilities of Deficiency

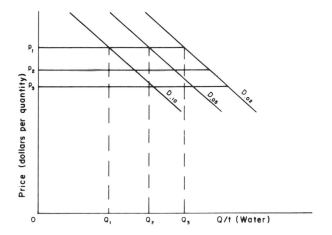

Figure 8. Demand for Municipal Water
at Different Probabilities of Deficiency

peaks are usually high. The expected annual damage due to both sources together amounts to the increase in annual benefits (ΔB) gained by investing more in the water utility plant; the expected annual cost of such an investment (ΔC) may then be compared with the benefits. Optimally investment should be increased until ΔB = ΔC. At present very little data is available regarding the losses resulting from the inability of waterworks to meet demands during periods of drought. For residential consumers in Massachusetts the reported losses are small: just over one dollar per head per annum.[57]

Summary of Planning Procedure

In a recent symposium on municipal water demand forecasting Howe proposed some needed improvements in the procedures of forecasting. He prefaced his analysis by making a basic assumption, viz., "urban development can be predicted independently of the costs and other characteristics of urban water supply."[58] Exception may be taken to this position. The economic growth of a region, expecially with respect to agricultural output dependent on irrigation, cannot be forecast without taking into account the 'requirements' and cost of water.

Judy observes in a critique on Howe's paper that "there is an apparently world-wide inverse correlation between aridity on the one hand and population density and value of economic activity on the other."[59] The economic and demographic future of a region is not independent of the availability of water supply; however, it is reasonable to assume that, given the other inputs required for city growth, the cost of water for municipal uses in general and residential uses in particular will not inhibit city growth and, therefore, Howe's basic assumption is acceptable with the qualification made above.

---

[57]C. S. Russell, The Definition and Measurement of Drought Losses: The Northeast Drought of 1962-1966, Resources for the Future Reprint No. 77 (Washington, DC: RFF, 1969) pp. 628 and 633.

[58]"Municipal Water Demands," p. 61.

[59]R. W. Judy, "Municipal Water Demands: A Critique," in Forecasting the Demand for Water, eds. W. R. D. Sewell and B. T. Bower, p. 83.

Howe recommends a sequence of steps for forecasting municipal water demand; it may be adapted to residential water use
forecasting:

1. Forecast level of population or number of dwelling units.
2. Using historical coefficients of water use, make a preliminary
estimate of water demands.
3. Preliminary water system design and costing.
4. Set a pricing schedule that reflects costs (or some other
preferred policy).
5. Revise the rate of use given the price structure determined
in step 4 (i.e., determine rate of use, given price).
6. Repeat steps 3 through 5 until demand is set equal to supply
on the target date.

The review and theoretical discussion of residential water
demand projections above provide the basis for improving the
present practice and the following sequence of steps may be
recommended for regional forecasts (Figure 9):

1. Forecast the   (a)   total population
                 (b)   number of single dwelling units (i)
                      metered individually or (ii) metered
                      as a group
                 (c)   number of dwelling units in high rise blocks
                 (d)   expected number of persons per dwelling
                      unit of type (b) (i), (b) (ii) and (c)
                 (e)   income level and change in income level
                      or value of residence of the subgroups as
                      defined above.
2. Find the historical coefficients of water use with respect to
type of dwelling, number of persons in residence and income
level.
3. Make a preliminary estimate of water demands in two steps:
(a) project water use cross-sectionally, (b) apply any time-
series information that may be available e.g., income elasticity,
change in taste and in technology and in efficiency in water use.
4. Preliminary system design and costing at some arbitrary
level of probability of deficiency (say 0.05).
5. Set a pricing schedule that reflects costs (or some other
preferred policy.
6. From price-quantity functions, determine the demand for

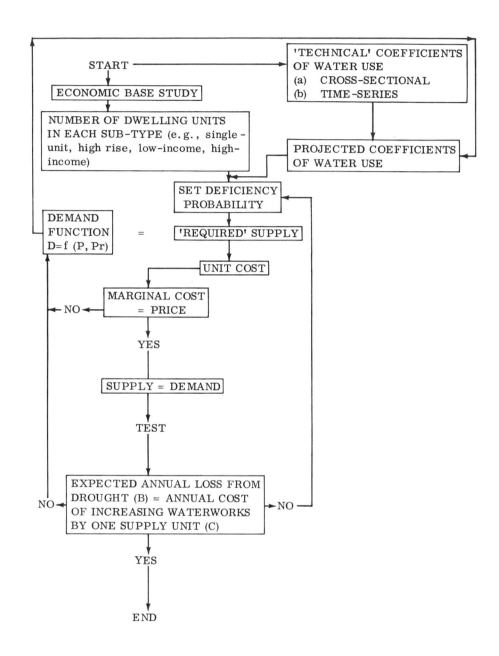

Figure 9. Residential Water Supply Programming

water given the price schedule set out in step 5.

7. Repeat steps 4 through 6 until the market is cleared at some acceptable price on the target date.

8. Determine whether the probability of deficiency is too high or too low. If the expected annual costs to the community of a water deficiency are greater than the required extra cost to obviate the drought, then the designed waterworks are too small. Conversely the probability of deficiency may have been set too high and the waterworks as tentatively designed are too large.

9. Repeat steps 4 through 8 until the cost of water supply and the size of the plant are determined and the forecast is internally consistent.

POLICY ALTERNATIVES: A FRAMEWORK FOR ACTION

The management of urban water resources involves the development of a system of collection, treatment, and distribution of water in such a manner that supply meets demand with the constraint that investment funds are limited and are a burden on the community (i.e., the water is not costless).

There are several policy alternatives that management can adopt in order to reach the objectives set out above. In Table 17 these are grouped according to the degree of change required from present practices. Ideally all the alternatives theoretically possible should be examined. The alternatives that are looked at first — and often last — are those that increase supply through further investment (pipelines storage, etc.). The management or engineering of water resources on the supply side is universally accepted. The management or engineering of water use is not so widely accepted. The alternatives listed in group B and C of Table 17 are feasible policies but, with the exception of regulatory police-action in times of emergency, these alternatives are not universal by any means and only metering is widely accepted. The effects of metering are most invariably off-set by faulty pricing schedules which reduce the effect of the marginal price to insignificance.

Table 18 gives some idea of the effectiveness of some policy alternatives in the Northeastern Illinois Metropolitan Area. The reduction from public water supply systems amounts to 177 MGD

TABLE 17. POLICY ALTERNATIVES IN RESIDENTIAL WATER-USE MANAGEMENT

A. Alternatives that do not require changes in policy-making attitudes:

   1. Increasing the supply of residential water through further investment:
      a. Increase treatment facilities,
      b. Increase storage facilities,
      c. Build pipelines,
      d. Desalination or some other method of water reclamation,
      e. Weather modification,
      f. Developing new ground water sources.
   2. Increasing re-use by means of a higher level of treatment of effluent.
   3. Waste water reclamation.
   4. Ground water recharge.

B. Alternatives that require some change in policy-making attitudes:

   1. Shifting water resources from some uses (e.g., agriculture or effluent dilution).
   2. Conserving water:
      a. Metering and pricing,
      b. Education and exhortation
         i. Inspection of house plumbing systems to check leakage,
        ii. Forecasting sprinkling needs over radio to avoid over-sprinkling
          especially in non-metered areas,
        iii. Encouraging rock gardens or other substitutes for lawns,
      c. Legislative action (e.g., reduction in household plumbing fixtures),
      d. Decreasing waste from the distribution system (e.g., mains leakage).
   3. Reducing the daily and seasonal peaks:
      a. Regulating time and day of sprinkling,
      b. Decreasing pressure at certain times,
      c. Levying seasonal charges.
   4. Reducing excess capacity by co-operating with neighbouring municipalities
      (e.g., selling water when the supply is greater than demand and buying water
      when the demand is greater).

C. 'Structural' or far-reaching policy alternatives:

   1. Accepting higher probabilities of deficiency during peak demand periods of short
      duration.
   2. Discouraging excessive use by means of prohibitory prices at the margin.
   3. Adopting a hierarchy of water quality (e.g., using a double plumbing system so
      that water of lesser quality is used for sprinkling, toilet flushing, etc.).
   4. Encouraging artificial lawns.

or about one sixth of the total daily pumpage from public systems
(1174 MGD).[60] It is clear that the demand for residential water
may be reduced substantially by means of policy decisions without
reducing the quality of the service.

[60]Northeastern Illinois Planning Commission, The Water Resource in Northeastern
Illinois: Planning Its Use (Chicago: NIPC, 1966) pp. 59, 117, 120.

TABLE 18. OPPORTUNITIES FOR REDUCING WITHDRAWAL
FROM DRINKING WATER SOURCES (Million Gallons Daily)

Management Measure	Reduction in Current Use 1961 (Million Gallons Daily)
1. Reduce underground leakage by 50%	115
2. Convert to new (smaller plumbing fixtures)	50
3. Eliminate over-irrigation of lawns	12
	177

SOURCES: Northeastern Illinois Planning Commission, The Water Resource in Northeastern Illinois: Planning Its Use, prepared by J. R. Sheaffer and A. J. Zeizel (Chicago: NIPC, 1966), pp. 117-120.

In conclusion it is useful to review the discussion so far. The adoption of a planning framework that allows for the examination and assessment of alternatives in residential water resources development stems from the need to examine ways of reducing the increasing claims on scarce public funds. Findings in this field may be applicable to other services requiring the allocation of public funds (e.g., education, highways). There are enough reported findings to enable management to design policies aimed at reducing demands for residential water.

The formulation and testing of a non-deterministic 'model' of residential water use (Chapter III) draws on previous work and the critical discussion on the factors determining residential water use. The assessment of some of the policy alternatives listed above is the focus of Chapter IV.

III

Structural relationships of residential water use: Theory and analysis

The theoretical and empirical work described in this chapter is designed to evaluate the hypothesis that residential water use varies with a number of factors such as the size of family, income, and price. The theoretical equation is fitted to sample data obtained in the study area and the coefficients so derived will be applied to the policy alternatives discussed in the next chapter.

Six variations of the model are formulated; the residential water use in gallons/day/dwelling unit averaged (1) over the year, (2) over the summer, and (3) over the winter period, are the three dependent variables used in turn with observations in two sub-samples viz. (a) single-unit residences and (b) groups of townhouses.

The differences in the level of water use between townhouses and single-unit residences will be ascribed to metering since this is the only identifiable difference between the two sub-samples;[1] in other words the marginal price for residential

---

[1]During summer the smaller lot of the townhouses is relevant too.

water use in townhouses is zero, as it is in single-unit residences in unmetered municipalities, with the important practical difference that it is possible to obtain data about residential water use for groups of townhouses but not for the unmetered single-unit residences.

STRUCTURE OF THE PROPOSED MODEL

Basic Elements of Residential Water Use

In residences having multiple-tap connections, the elementary units of residential water use are the water using appliances in and around the house: taps, baths, toilets, water-hoses, sprinklers, lawn, garden, car, swimming pool, air conditioning units, dishwasher, garbolator, and so on.[2]

The amount of water used by an appliance i in a given time period may be stated as:

$$d_i \equiv n_i \, w_i \, p_i \qquad (i = 2, 2, \ldots, m) \qquad \text{where}$$

n_i stands for the number of appliances of type i,

p_i stands for the probability of the appliance being turned on and

w_i stands for the rate of water use by appliance i.

If there are m types of appliances and each has a uniform rate of use, one could calculate the amount of water used by a household (D) at a given time by summing d_i

$$D \equiv \sum_{i=1}^{m} d_i \equiv \qquad n_i \, w_i \, p_i \qquad \qquad (9)$$

This simple identity describes the process of water use by a household and the formulated model has to be compatible with the propositions that may be derived from the identity. They

---

[2] The minimal amount of water required to satisfy a person's physiological needs amounts to about 1 litre/capita/day according to Bernd H. Dieterich and J. M. Henderson, Urban Water Supply Conditions and Needs in Seventy-Five Developing Countries, World Health Organization Public Health Papers No. 23 (Geneva: WHO, 1963), p. 27. In high-income countries the water required by each household to meet these physiological needs may be considered negligible.

are stated in summary form below and will be taken up again
in the discussion of the estimating equation.

1. The amount of water used by a household at a given time is
the direct result of the consumer's ability and willingness to
purchase and use household goods such as baths, sinks, showers,
and garden space. Residential water use is complementary to
other household activities.
2. Residential water demand is a composite demand. The total
demand is the sum of the water used to complement activities
such as gardening, washing, and waste disposal. In addition, it
is useful to note that there is a hierarchy of water-complementary
activities. The consumer would give up the least preferred water-
complementary activity first if he had to.

   Total prohibition and physical restrictions are, therefore,
particularly unsuited to residential water use management because they deprive all consumers of all water-complementary
activities, irrespective of their place in the hierarchy of
perceived needs. On the other hand, pricing policies respect
the consumer's right to choose to use water for some purpose
and not for others.
3. The composite nature of residential water use makes the
demand for this commodity vary with the time of day and year,
since the number of water using appliances and the frequency of
their use increases during periods of hot dry weather. This
means that various measures of D are available. The seasonal
variations in demand are considered critical in the design of
water supply systems and a measure of the summer rate of use
is a relevant variable in a policy-oriented model.
4. The identity (9) is a measure of how much water is used by a
household during a given time interval. Even if one had complete
knowledge of the present and projected values of m, n_i, w_i, and
P_i for all households in a community, one would still be no
nearer to understanding the why of the residential water use
phenomenon. The object of the estimating equations is to reduce
the amount of information required to forecast D by discovering
a small number of 'explanatory' variables.

   Only those variables that are known to affect m, p_i, w_i, and
n_i in (9) above can be said to enter into the structural relationship

of water use.[3] Empirical relationships — even if statistically
significant — that are not compatible with the basic identity may
be misleading. An example already discussed is the inverse
relationship between the level of residential water use and the
average price of water. Price can affect D only to the extent that it
dissuades the consumer from a water-complementary activity and
the only relevant measure of price is the marginal or variable or
commodity charge.

5. The composite and complementary nature of residential
water use results in gradual changes in the individual household's patterns of water use over time since the purchase of
semi-durable water-using appliances does not adjust instan-
taneously to changes in price or income or technology.

6. There are innumerable variables that affect the magnitude
of m, p_i, w_i, and n_i, in (9) above. Some of these variables
may be ascertained by means of a detailed household survey
(e.g., duration of family holidays, frequency of watering garden
or lawn, use of a backyard rink in winter). Others would pass
unnoticed even then (for example, undetected leaks in the household plumbing system). One should, therefore, expect large
errors from the estimating equation; rather than add more
variables in order to decrease the errors from the equation, it
would be more efficient to concentrate on the significance—or
otherwise—of variables that are relevant to policy making
(e.g., metering, price).

A measure of the income level of the household is one of the
'explanatory' variables since the variety (m) and number (n_i)
of water using gadgets around the house reflects the ability of
the household to pay for such appliances.

The probability that an appliance is used (p_i) depends partly
upon the income level, partly on the number of persons in the
household, and partly on the price of water.

The rate of water use by each appliance (w_i) is dependent
upon its technical efficiency. In the case of residential water

[3]The structural relationships may include technical relationships (e.g., number of persons
in residence) and behavioural relationships (e.g., price and income effects). The terminology is adopted from Karl A. Fox, Intermediate Economic Statistics (New York: Wiley,
1968), p. 81 ff.

use this efficiency level is partly under the control of the user; care of use and the repair of water using appliances (particularly the household plumbing system) are relevant to the magnitude of w_i. The rational consumer adjusts to the marginal price of water by using water more effectively.

The discussion so far may be summarized as follows: income sets a certain level of water-complementary activities; the number of persons in residence has a technical relationship with water use; the pricing of water use enters into a behavioural relationship. It is required to formulate the hypothesized relationships in operational terms and to define the variables on both sides of the equation.

The Hypothesized Equation

It is hypothesized that:

$$D \equiv \begin{matrix} WUa \\ WUs \\ WUw \end{matrix} = f(V, \ L, \ Li, \ Np, \ P, \ Nb, \ A, \ F) \tag{10}$$

and specifically that:

$$D \equiv \begin{matrix} WUa \\ WUs \\ WUw \end{matrix} = b_0 + b_1 V + b_2 L + b_3 Li + b_4 Np + b_5 P + b_6 Nb + b_7 A + b_8 F + U \tag{11}$$

where the subscript (i = 1, 2, 3, ... n) referring to observations is omitted from all expressions and where

WUa	is the water use in gallons/day/dwelling unit (annual average);
WUs	is the water use in gallons/day/dwelling unit (summer period average);
WUw	is the water use in gallons/day/dwelling unit (winter period average);
V	is the assessed sales value of residence in hundreds of dollars;
L	is the lot size in hundreds of square feet;
Li	is the area not covered by buildings in hundreds of square feet;
Np	is the number of persons in the dwelling unit;
P	is the variable price of residential water in cents/1000 gallons;
Nb	is the number of billing periods;
A	is the number of gallons/day/dwelling unit that are allowed with the minimum bill;
F	is the fixed bill for one billing period in cents;
U	is an error from the equation having the usual stochastic characteristics.

The expected functional form of the relationship will be discussed below after the variables in the quation are defined more fully; but at first the linear functional form will be

80

assumed. The magnitude of the regression coefficients cannot be specified on a priori consideration but as is indicated below the sign of the coefficients of V, L, Li, Np, Nb, and A should be positive while the coefficients of p and F should be negative.

Independent Variables in the Equation

Price
In accord with generally accepted utility theory it is proposed that a consumer derives utility or satisfaction from the use of water in his residence so that one can describe the amount of water consumed to be a function of income, the price of water and the price of other items of consumption; in practice it is necessary to restrict price variables to those pertaining to the commodity itself and close substitutes. In the present instance all cross effects (i.e., the prices of related goods) can be neglected since there are hardly any close substitutes for residential water except care in use.

It is hypothesized that in general the effect of a higher variable charge for residential water is reflected in a reduction of water use through the efforts of the consumer to use water more efficiently or more carefully: leaks are repaired more promptly, more care and time is devoted to lawn watering; over a longer period of time water use patterns are changed by replacing high water-using home appliances with low water-using models or by reducing water-complementary activities (e.g., reducing the area in lawn). These differences due to 'habits' are the ones observed in cross-sectional data i.e., data obtained from points in space at a given time; a sample of cross-sectional data is used to validate the hypothesized structural relationship.

It may be argued that the cost of the water to the consumer is small compared to the total household budget and to the cost of the water-complementary appliances and that therefore the impact of price differences on water use by households is negligible. Inquiries made at the water revenue offices of the several municipalities included in the sample showed that water bills are queried by about 5 to 10 per cent of the consumers. The customers who complain are often new residents who used to live in unmetered municipalities or they have faulty equipment in the house or garden. The rate of inquiries about water bills shows that at least some consumers are responsive to the size of their water bill.

81

Defining the price of water

The price of water itself raises another problem. In the study area residential water is priced in two parts: a minimum bill which includes a quantity of water/day/dwelling unit that varies from 20 gallons in Newmarket to 67 in most municipalities and to over 100 gallons in Ajax. Water consumption in excess of this amount is charged at the 'commodity' price which will be referred to as the variable or marginal price.

The amount of water allowed with the minimum bill is usually small enough to be exceeded by almost all consumers. Municipalities where this amount of water is greater than 10,000 gallons/ 3 months (or 111.1 gallons/day/dwelling unit) were excluded from the sample because in these municipalities the variable price for relatively low water consumers will be zero. Barrie and Ajax are excluded on these grounds.[4]

In two municipalities (North York and Markham Township) the variable price of water has a declining step which falls within the normal range of water consumption by residences; this means that in these municipalities the price paid for the last gallon(s) of water is a declining function of the quantity taken. As it was pointed out above the hypothesized function

WU = f(variable price)

is stated in such a way that it cannot be rejected — low water use would necessarily be associated with high prices and high water use would be associated with low prices. These municipalities are excluded from the sample.

In the present study the units of observation are individually metered consumers and, therefore, it is possible to ascribe the variable price that was actually paid in 1967.[5] The use of average

[4]In the sample some consumers have minimum bills; it is intended to keep these observations in the sample for several reasons: (1) these consumers either have a low preference for water-complimentary activities or they may be trying to avoid paying more than the minimum bill; this means that the variable price equals or exceeds the marginal utility of water for a consumer on a minimum bill and therefore the price acts as a constraint on water use; (2) it may also be that the low water use in 1967 was a matter of chance; (3) low water use is usually associated with older persons living on fixed incomes in low-assessment residences and the inclusion of such observations adds to the range of the conditions to which the estimating equation applies.

[5]In North York, Scarborough and York there were price changes in mid-1967 and these municipalities were excluded from the sample.

82

prices was questioned in the critique of previous studies. It was argued that in such studies (e.g., Fourt, Howe and Linaweaver, Haver and Winter, Turnovsky) the finding that water use is a declining function of 'price' is open to two interpretations: (1) it may be a spurious relationship or (2) it may reflect consumer behaviour if the average price reflects the marginal price charged to the consumer. This assumption was tested in the study area: for the six municipalities included in the sample (Etobicoke, East York, Mississauga, Newmarket, Burlington, and Pickering Village) the simple correlation between the average cost of municipal water and marginal price is 0.4. [6]

Price is the most important policy variable and careful definition is essential. Howe and Linaweaver argue that "in studying the impact of price on water demand, the proper concept of price is defined by answering the question 'What charges can be avoided or changed in magnitude by the decision being made by the decision-making unit?'"[7] In this study the price of water ascribed to the individual residential consumer is the net charge in cents/1000 gallons that he pays or would pay beyond the quantity allowed with the minimum bill. This is the price to which a logical consumer is sensitive. [8]

In a few municipalities (e.g., Newmarket, Pickering Township) sewerage charges are included in the price of water. The full charge is correctly ascribed to the individual residential consumer since he has the option to use less water and save the total charge for water and sewage. For the same reason the net price is applied, i.e., the discount for prompt payment is allowed and any penalty for late payment is not added.

[6]Among the six items are two 'extreme' examples viz., Newmarket (where low acreage cost is paired with the high marginal price) and Pickering Village (where high average cost is paired with a high marginal price). When both are omitted the correlation coefficient is less than 0.1. With ten observations (i.e., the six municipalities in the sample and the neighbouring municipalities of North York, York, Scarborough, and Pickering Township) the value of r is 0.4.

[7]Charles W. Howe and F.P. Linaweaver, Jr., "The Impact of Price on Residential Water Demand and Its Relation to System Design and Price Structure," Water Resources Research 3, no. 1 (1967):14.

[8]Prices are abstracted from: "1967 Annual Survey of Municipal Water Rates of Ontario," Water Works Digest (Hamilton, Ontario: Stanton Pipes Ltd., published annually).

Assessed sales value of residence

It is commonly accepted that the income of the consumer affects consumption patterns; with respect to residential water use the level of income affects the variety, number and frequency of use of water-complementary appliances. A variable that represents the income level of the household has been included in equation (11) the assessed sales value of the residence of the consumer is substituted for the measure of income flow both on a priori and practical grounds.

The assessed value of a residence — as well as the lot size and the number of residents — is available from the assessment rolls or the population census kept for assessment purposes. The assessment figures for the various municipalities can be reduced to a common base by using the Assessment Equalization Factors determined by the Assessment Branch of the Department of Municipal Affairs of the Province of Ontario for 1968 assessment.[9] The resulting assessed sales value approximates the value of and is related to the physical conditions of the residence.

In assessing a house, the number of bathrooms, the garage, the lot size and other physical assets of the residence are taken into account. Since water is used to complement the use of appliances such as, sprinklers, baths, sinks, washing machines, swimming pools, lawn, etc., some measure of real estate is a more relevant explanatory variable for water use than some measure of income flow. In other words water use is related to income as reflected in the amount of real estate held by the consumer. Some earlier studies of residential water use which used a similar measure of income level were discussed previously.

Assessed sales value of residence reflects the presence of home improvements that are usually water-complementary (baths, lawn, swimming pool); in addition this variable may be considered as a surrogate for income assuming that high income consumers live in high assessment residences.

Apart from the difficulties of appraising property in order to arrive at an "opinion of value that should be supported by factual

---

[9]Mr. Bailey of the Assessment Branch has kindly explained how the Equalization Factor is derived. It is the median value of the Assessment/Sales ratio (%). The significance of this median value is measured by means of the Coefficient of Dispersion (CD) where CD = average deviation from equilization factor divided by the equalization factor. If the CD is 20% or less the Assessment/Sales ratio median is accepted as the Equalization Factor.

data, "[10] there are two further observations that should be made about this variable. The figures for assessed sales value have to be comparable not only from residence to residence in the same municipality but also from municipality to municipality.

Young, in a background study for the Smith Report on Taxation, has compiled data that demonstrate the reliability of comparisons of assessment in the same municipality. [11] Young has used the estimating equation:

$$\text{Assessed value} = b \text{ (Sales value)}$$

and found that the standard error of the estimate, as a percentage, was less than one per cent in 13 municipalities and was less than 2 per cent in all but one municipality in a sample of 22.

The assessment/sales ratio—the B coefficient in the above equation—varies from 68 per cent in Sarnia to 21 per cent in Woolwich. Therefore inter-municipal comparisons are valid to the extent that the correct assessment/sales ratio has been arrived at. There is an element of 'chance' in the selection of data because the observations are the actual sales which have occurred during the year.

Since the assessed values reflect sales values, the equalization factor reflects inter-municipal differences on the whole but it is possible that the equalization factor for one particular municipality may be too high or too low. This may create difficulties in the interpretation of the estimating equation of water use.

If the equalization factor is too high (i.e., the estimated sales value is too low) then the errors from the equation will be consistently positive and vice-versa. In other words the residuals for that municipality will show a systematic pattern and it may be difficult to decide whether it is the pattern of water use that is different or whether the systematic bias in the assessed sales value is the real cause of the pattern in the residuals.

---

[10] These problems are discussed in Ontario Department of Municipal Affairs, Manual of Assessment Values, 2nd ed. (Toronto: Queen's Printer, 1954), and Appraisal Notes for the Assessor (Toronto: Queen's Printer, 1964).

[11] G. I. M. Young, "A Study of the Assessment of Real Property in Ontario," prepared for the Ontario Committee on Taxation, 1964. Mimeographed.

Lot size and land not covered by buildings
The size of the lot of a residence may affect the use of water in two ways: (1) it reflects the general living standards and style of life, and (2) it reflects the sprinkling requirements during the summer; the sprinkling requirements not only increase the average use of water but contribute to the summer peak.

The method of appraising property for tax assessment takes the size of the lot into account. As one would expect, the lot size and the assessed value of residence are positively correlated and either variable may be dropped from the equation.

For the first 76 observations data were available on the area of land covered by buildings.[12] By subtracting from the lot size one can obtain the area of land not covered by buildings and possibly irrigated during the summer. As may be expected this is highly correlated with lot size and assessed sales value. For technical reasons data about this variable were not available after the first 76 observations. Results from the first 76 observations suggest that this variable is correlated with size of lot and may be dropped from the equation (Table 19).

The regression equation of Li on V and L is:

$$Li = -0.5 + 0.01V + 0.71L \qquad (12)$$
$$(0.01) \quad (0.05)$$

$R = 0.87 \qquad S.E. = 10.64 \qquad F \text{ value} = 119.9$

It will be recalled that Howe and Linaweaver included a similar variable, viz., irrigable area per dwelling unit. The coefficient for this variable was unstable and the authors suggested that the instability may be "due to the small range of variation" in the observations; high intercorrelation with the value of residence may be another explanation as the present results indicate.

Number of persons in the dwelling unit
This should be the most significant variable. Over most of the year the greatest use of residential water is for bathroom purposes and this particular use of water is some linear function of the number of persons in the residence. For example, the

---

[12]Mean = 34.45 and standard deviation = 21.63 both in hundreds of square feet.

TABLE 19. SIMPLE CORRELATION COEFFICIENTS
OF ASSESSED SALES VALUE OF RESIDENCE, LOT
SIZE AND AREA NOT COVERED BY BUILDINGS

	V	L	Li
V	1.00		
L	0.697	1.00	
Li	0.744	0.842	1.00

NOTES: V   is the assessed sales value of residence
            L   is the lot size
            Li is the area not covered by buildings

average toilet tank holds 4 gallons of water, the average sink
about 1 gallon and it takes about 10-20 gallons for a bath or
shower. Each person is likely to use about 30 to 40 gallons
daily for such purposes.

It is therefore somewhat surprising that the results reported
in the literature are not clear-cut. The explanation suggested
in the critique is the lack of variance in the variable Np; if the
preliminary sample of observations does not give significant
coefficients, stratification of the sample with respect to this
important variable may be required.

It would be helpful if the number of persons in the household
may be divided into adults and children and into those who spend
most of the day at work and those who stay at home; information
regarding the number of days the family spends away from home
during summer and winter would be useful too but none of these
data were available.

Number of billings

If the bill acts as a deterrent to the careless use of water, then
the number of billing periods and the amount of water allowed
with the minimum bill could be expected similarly to affect water
use.

The number of billing periods in the study area is either 3 or
4 or 6 and there may not be enough variance to provide a proper
test of the hypothesis, viz., since the water bills are not pro-
hibitively high (about $3 to $5 a month) the more often the billing
the lesser is the impact. If the bills were very high then the

effect of the more frequent billing would be to reinforce the effect of the high bill i. e., to reduce the amount of water use.

This variable (like price, the size of fixed bill and the amount of water allowed with the minimum bill) may act as a 'dummy' variable that differentiates among the municipalities. In other words, it may be possible that this variable may be accounting for some variability in water use that is due to some condition peculiar to the municipality which is correlated with this variable. Physical restrictions on water use is an example of such a peculiar condition and care was taken not to include municipalities which had restrictions on water use in 1967. Other causes are not so apparent (e. g., a higher preference for greener lawns or systematic under-assessment of property).

Amount of water allowed with the minimum bill

An amount of water is allowed with the minimum bill, often at an average price higher than the commodity charge (i. e., the marginal price of water). This has implications for policy: the abolition of the minimum bill would provide the consumer with an incentive to reduce water use from the first unit of consumption. At present there is an incentive to reduce water use after the allowance with the minimum bill is used. As it was pointed out above, the allowance may be 100 gallons/day/ dwelling unit in some municipalities.

The demand curve of a consumer for residential water (D) is shown to cut the quantity axis (Q/t) indicating satiation at zero price (q_2) (Figure 10). The demand curve probably approaches the price axis asymptotically because a minimum water service is demanded even at a high price; this part of the demand curve is shown by a dashed line.

At price p_1 the consumer buys q_1 of water. If the amount of water allowed with the minimum bill (A) is small $(A_1 \leqslant q_1)$ the consumer would still buy q_1. If $q_1 < A_2 \leqslant q_2$ the consumer could use A_2. One should, therefore, expect the water use by households to increase as the allowance of water with the minimum bill increases.

Size of minimum bill

It is hypothesized that the amount of water used by a household is responsive to the size of the bill; on the whole the careless use of water is curtailed but if the bill is high enough it may be an

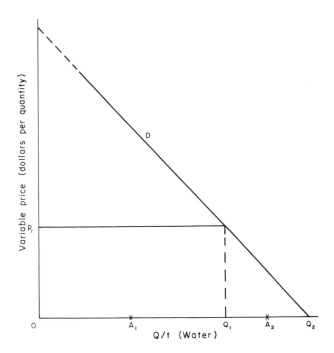

Figure 10. Effect on Residential Water Use of the Amount of Water Allowed with the Minimum Bill

incentive to reduce some water-complementary activities.

The water bill consists of a variable bill and the fixed or minimum bill. The variable price of water covers one aspect of the 'billing incentive': it is proposed that the size of the fixed bill, irrespective of the length of the billing period, be used as a measure of the impact of the bill as a whole on the consumer. The rationale from the consumer's view is "the water bill is high enough, so let's not increase it any further than is necessary." The relationship of this variable with water use is expected to be negative.

The interpretation of this variable is not clear cut. Like the number of billing periods, the variable price of water, and the

89

amount of water allowed with the minimum bill, this variable may act as a 'dummy' i.e., it may reflect related conditions pertaining to a municipality as a whole (e.g., over-assessment) rather than the impact of the size of bill. The distribution of the residential errors should indicate whether the variable is acting as a dummy or not.

Dependent Variables in the Equation

It is important to define the dependent variables particularly because there are 'errors of measurement' that increase the standard error from the equation. The errors of measurement are caused by lack of homogeneity in collecting and keeping records of water use among the municipalities.

Some meters (e.g., in Etobicoke) are read to the nearest 1000 gallons and therefore they may be an error of 5.5 gallons/ day/dwelling unit in each observation. This error is relatively small because the water use is over 100 gallons/day/dwelling unit. In any case the data cannot be improved and these errors should be randomly distributed.

It is impossible to include the same days of the year for all observations; with regard to average use over a year (WUa) this problem is of negligible importance but it makes the values of the average summer (WUs) and average summer use (WUw) subject to further random error. Reading dates vary from district to district and billing periods need not — and often do not — correspond to the seasons of the year. For example, some summer days may be included with the spring or the autumn period.

Since it is desirable to compare summer and winter use, a clear operational definition is required:

1. When the number of billing periods is 4:
   a) Summer use is defined as the average use of the period which includes 62 or more days of the 123 days in May through August;
   b) Winter use is defined as the average use of the period which includes 68 or more of the 135 days from November 1 through March 15.
2. When the number of billing periods is 6 (e.g., Burlington, Newmarket):
   a) Summer use is obtained by adding the consumption of the two billing periods from May through August and dividing by the number of days;

b) Winter use is computed from the billing period that includes 30 or more of the 90 days from December through February.

The average water use over a year is computed from the total of the 4 or 6 billing periods; all have about 365 days and run from about January to December.

The Functional Form of the Estimating Equation

The first step in specifying a model and fitting it to data is the choice of variables and the second step is to define the variables operationally. The third step is to specify the functional form of the relationship.

Howe and Linaweaver state that theoretical considerations fail to specify a unique functional form;[13] both linear and multiplicative forms were fitted in their study but not all the empirical results were reported. For domestic demand the results show that both forms of the equation give comparable results; for sprinkling demand only the multiplicative form is reported. [14]

Turnovsky follows tradition and assumes that "the demand relation to be estimated is indeed linear. If for no other reason, this can be justified as an approximation of more general functions in the neighbourhood of the explanatory variables." [15]

While there are strong theoretical indications of a curvilinear relationship, in the light of the discussion of previous models, the linear equations would be interesting initial output. Since the standard error from the equation is likely to be great, curvilinearity in the relationship may be easily masked. The assumption of a linear relationship is a strong one and there are theoretical reasons for believing that the relationship is curvilinear. Some hypothesized curvilinearities are set out below and both linear and curvilinear forms of the equation will be tested against sample data.

[13]"The Impact of Price on Residential Water Demand," p. 20.

[14]Ibid., p. 24.

[15]S. J. Turnovsky, "The Demand for Water: Some Empirical Evidence on Consumers' Response to a Commodity Uncertain in Supply," Water Resources Research 5, no. 2 (1969):351.

The independent variables measuring income level (V) and the number of persons in residence (Np) should show some decline in the slope as they increase. As the assessed sales value of residence increases the use of water may be expected to increase proportionally at first but beyond some point the use of water will not increase as fast. For example, water uses for personal hygiene and car washing do not increase indefinitely with income. There are some water uses that may decline with income (e. g., a high-income family takes longer holidays as income rises).

The same logic applies to the number of persons in residence. In the house there are some uses of water that do not depend on the size of family (e. g., lawn-sprinkling, leakages). As the number .of persons in the dwelling unit increase such water uses are averaged over a larger number of persons.

The functional form of the relationship between price and residential water use is the one most relevant to management. Even if the linear form holds good for individual water uses, the demand function for total water use by a household should be curvilinear. This follows from the nature of the basic elements of residential water use.

In Figure 11(a) the demand shown as d_1 represents demand for essential water uses such as drinking, cooking, washing clothes, personal hygiene, waste removal; the consumer is willing to pay a high price for water for such purposes and at zero price the demand is still small. The curve d_2 denotes demand for water of lesser importance to the household (e. g., for lawn-watering, dishwasher, garbolator), while d_3 indicates the demand for water of least importance (leakages, careless use in sprinkling, 'waste'). The rate of decrease with price is least for d_1 and greatest for d_3. The three demands are summed horizontally in Figure 11(b) and the total demand curve must be curvilinear. [16]

If the regression coefficient of price (b) is not significantly different from zero, the hypothesis that water use is independent of price cannot be rejected. If the regression coefficient of price

---

[16]The slopes of d_1, d_2, and d_3 need not be different for this relationship to hold. However, the magnitude of the slope that reflects the perceived hierarchy of water uses is attractive theoretically and is suggested by some of the reported empirical work in chapter II above, e. g., domestic use is less responsive to metering and price than sprinkling use.

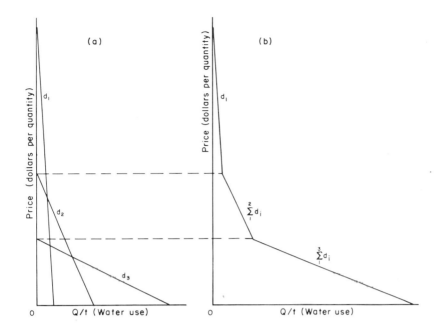

Figure 11. Demand for Residential Water

with respect to summer use (i.e., b_s or the slope of $\overset{3}{\underset{1}{\Sigma}} d_i$) is greater than the coefficient for winter use (i.e., b_w or the slope of $\overset{2}{\underset{1}{\Sigma}} d_i$) then curvilinearity is indicated.

The implications for management are clear: when price is low the inclusion of low-preference water uses makes water use at that point of the demand function highly responsive to price change; when price is high the least-valued water uses (e.g., waste) are curtailed making water use at the point relatively unresponsive to price change. In municipalities where the price is high, the effect of a further increase is much lower in absolute terms than the effect of an equal price increase in a municipality where the price is low.

Estimating the Parameters

Two types of dwellings will be considered: single-unit or low-density housing which are metered and groups of townhouses which are metered as a group. For each type of dwelling the objectives of the sampling design and the procedure of drawing the sample(s) will be described before passing on to the discussion of the estimating equations.

RESIDENTIAL WATER USE IN
METERED SINGLE-UNIT DWELLINGS

The Sampling Design

Sampling objectives
The object of the sampling procedure described below is to obtain a sample of observations of metered water use in single-unit dwellings in 1967. These dwellings are drawn from those municipalities in the Toronto Region in which residential water meters are read every two to four months. Each observation in the sample represents approximately 8000 persons or 3000 dwellings. The items in the sample are to be used in multiple regression analysis with a view to obtaining the behavioural and other structural relationships between water use in single-unit dwellings and postulated explanatory variables.

The adoption of a sampling design was influenced by both the practical considerations of expense in time and effort in obtaining the data and the analytical procedures to which the data were to be subjected. Before describing the procedure for drawing the items in the sample some of considerations will be set out.

The main aim of this study is to derive accurate regression coefficients of residential water use with respect to some of the postulated explanatory variables; these coefficients will be used in the discussion of the alternative policies. It is also intended to examine the extent to which the independent variables 'explain' the large variances that are known to occur in residential water use. These objectives can be efficiently attained by sampling among the residential water users in the study area.

In multiple regression and correlation analysis the investigator is interested in the regression coefficients and their standard errors; the coefficient of multiple correlation which gives a summary measure of the total variance accounted for by the linear or curvilinear relation fitted; the partial correlation coefficients which indicate the proportion of variance explained by each independent variable after the other independent variables have accounted for their share of the explanation; the F-statistic which indicates the significance of the regression as a whole; and the standard error from the regression plane.

The multiple, simple, and partial correlation coefficients and the standard error of the regression coefficients are sensitive to changes in the standard deviation of the independent variables since Sx_i appears in the denominator of the formulae for these statistics. The sample correlation coefficients may be regarded as estimates of the corresponding parameters without qualification only when the sampling is random from normal multivariate distributions.[17] Kendall argues that the assumption of normality in the parent populations may be relaxed.[18]

Only one of the samples described below is completely random and can, therefore, be interpreted without qualification with respect to the simple, partial, and multiple correlation coefficients. In the other samples the main purpose is to obtain accurate regression coefficients and to examine the predictive value of specific independent variables such as price, the number of billings per year, the fixed charges per bill and other independent variables relevant to policy changes in residential water use management. These variables require stratification and under such circumstances selection with respect to the X-variable is not only inevitable but also desirable.

[17]M. Ezekiel and K.A. Fox, Methods of Correlation and Regression Analysis: Linear and Curvilinear (New York: Wiley, 1959), pp. 279-281; and K.A. Fox, Intermediate Economic Statistics (New York: Wiley, 1968), pp. 12, 185 have useful discussions of this topic.

[18]M.G. Kendall, "Regression, Structure, and Functional Relationship, Part I," Biometrika 38 (1951):11-25.

If we are interested in obtaining quite accurate estimates of regression coefficients, and if it is possible to subject the independent variable to experimental control, one can obtain small standard errors with a modest number of observations provided that the observations are thinly spread out over a wide range. Under such conditions and supposing there also exists a natural population of values of the independent variable from which we could sample at random, the random sample will inevitably provide us with a large number of values relatively close to the mean and contributing relatively little to the standard deviation of the independent variable.

Random sampling when controlled experimentation is possible, is quite inefficient in terms of information gleaned per observation. If the observations can be obtained by both random and controlled experimental methods, the designed experiment should be used.[19]

Sampling procedures

Sample I — With a sampling ratio of one observation per 8000 population only one or two observations were likely to be drawn from the majority of the municipalities in the Metropolitan Toronto and Region Transportation Study (MTARTS) area. A pilot study indicated that the task of obtaining a few items from a large number of municipalities was unduly expensive in time, effort and money. In addition, the quality of the data in the small municipalities was lower than that obtained from the larger municipalities; for example, in some municipalities (Pickering Village, Pickering Township, Markham Township) water meters are read by the customers and the date of reading could not be accurately determined.

On the other hand, there is no reason to believe that the conditions affecting residential water use (e.g., prices, assessed value of residence and family size) vary systematically between small municipalities and larger municipalities. It seemed best to spread the number of items in the sample over the larger municipalities, viz., Etobicoke (population 266, 458), East York (97, 555), and Mississauga (105, 000), considering them as one cluster.[20] The items in the sample of households

---

[19] Fox, Intermediate Economic Statistics, p. 223. See also E. J. Williams, Regression Analysis (New York: Wiley, 1959), p. 22.

[20] Originally the 'cluster' of large municipalities included Scarborough, York, and North York. Observations from these 3 municipalities were discarded because of the price change during 1967.

were selected in two stages. Cumulative numbers in proportion to the population were assigned to each of the three municipalities and random numbers were drawn until 67 items were allotted. The second stage of the sampling procedure invblved picking at random the required number of accounts allotted to each municipality.

The spread or variation of the X-variables in Sample I and in the following samples is summarized in Table 20. The number of billing periods per year shows no variation at all; the critical variable from the point of view of policy, price, varies from 39 to 45 cents/1000 gallons. It is hypothesized that even under the present pricing policies, marginal (or commodity) price explains some of the variance in residential water use; with rising prices and/or a more rational pricing system the marginal charge for residential water use would become even more important as a policy variable. It is considered important to obtain accurate estimates of the regression coefficients with respect to price and therefore to increase the range of the price of water. Sample II — In order to extend the price range two smaller municipalities in the MTARTS area were included as clusters in the enlarged sample II, viz., Burlington and Newmarket where the price of residential water was 30 and 80 cents/1000 gallons respectively: 12 items were assigned to each municipality.

In this sample price is used as a stratification factor in order to ensure that the observations are spread out over the range of prices in the study area. This sample is random for all variables except price; with respect to price the bulk of the observations were assigned to the middle price ranges since the items were already available from sample I. Selection within strata is made randomly but there is no reason to have a constant sampling fraction. [21]

In Kendall's terminology, the sample is partially determined and the rigorous interpretation of the correlation coefficients with respect to price is somewhat impaired. Provided price is not correlated with other independent variables (e.g., value of residence) one can make statistical inferences at the appropriate

---

[21]C.A. Moser, Survey Methods in Social Investigation (London: Heinemann, 1958), pp. 79 and 88.

TABLE 20. DISTRIBUTION OF OBSERVATIONS BY MUNICIPALITY, ASSESSED SALES VALUE OF RESIDENCES AND NUMBER OF PERSONS IN HOUSEHOLDS

Sample	Number of Observations	Municipality	Price/ 1000 Gallons	Nb	A	F	Assessed Sales Value of Residence ($ 000's)							Number of Persons in Household			
							5-10	-15	-20	-25	-30	-35	-45	1-2	3-4	5-6	6-8
	12	Newmarket	80	6	20	100	2	7	4	1	–	–	–	2	4	4	2
	25	East York	44	4	36	285	2	7	11	3	2	–	–	10	9	4	2
IIa II I	29	Etobicoke	39	4	66	300	–	7	12	6	2	2	–	6	15	6	2
	13	Mississauga	45	4	66	540	–	–	4	1	3	4	1	1	5	5	2
	12	Burlington	30	6	66	400	–	1	2	6	2	1	–	2	3	5	2
	11	Pickering V.	71.25	6	66	285	–	1	6	2	1	1	–	4	4	3	–
TOTAL							4	21	39	19	10	8	1	25	40	27	10

NOTES: Nb is the number of billing periods per annum

A is the number of gallons of water allowed with a minimum bill

F is the minimum bill in cents per billing period

confidence levels about the population correlation coefficients of the other variables.

Sample IIa — Eleven observations from Pickering Village are added to the 91 items in sample II. The price of residential water in Pickering Village is 71.25 cents/1000 gallons; this sample acts as a check on the regression coefficients with respect to price since there are more observations at a high price. Unfortunately the meters in Pickering Village are read by the consumers (except for an annual check reading); billing dates are, therefore, not available. The error in the length of the billing periods is small over a year but it may be considerable over a 2-month period. For this reason only the annual average water use was considered to be useful.

The Estimating Equation
The theoretical basis for the model has been set out in a previous section of this chapter. It is proposed to examine the equations as fitted to the samples of observations. Some of the postulated variables may not have significant regression coefficients; these will be briefly examined and an equation of 'best fit' will be selected; the selected equations will be used in the application of policy alternatives in the next chapter.

The linear equations for each sample are given first, followed by some non-linear equations. Then the assumptions with respect to the residuals are examined. The assumptions of least squares analysis may be briefly stated as follows:[22]

1. The independent or X-variables contain no random errors.
2. In the sample, the residuals from regression are uncorrelated with the independent variables.
3. In the sample the expected value of the residuals is zero; the variance of the residuals is assumed constant for all possible combinations of values of the independent variables; the successive residuals (when some logical basis for ordering exists) are uncorrelated.
4. A linear relationship exists between the dependent variables (Y) and the explanatory variables $X_1, X_2, \ldots X_k$ and a disturbance term (u) so that

$$Y_i = a + b_1 X_{1_i} + b_2 X_{2_i} + \ldots b_k X_{k_i} + u_i. \qquad (1 = 1 \text{ to } n)$$

5. No exact linear relationship exists between any of the X-variables.
6. The number of observations (n) exceeds the number of parameters to be estimated (k + 1).

Regression analysis is used in order to seek variables that 'explain' some of the variation in the dependent variable. In groups of observations relating to residential water use this variance is considerable. The independent variables 'explain' about half the variance of the water use/day/dwelling unit. The rest constitutes the error from the equation. It may be attributed to (1) errors in measurement of the dependent variable; (2) variation due to factors not taken into account in the regression, and (3) "a basic and unpredictable element of randomness in human responses."[23]

All three types of error lower the value of the multiple correlation coefficient and increase the standard error of the

[22] J. Johnston, Econometric Methods, International Students Edition (New York: McGraw-Hill, 1963), pp. 5-9, pp. 106 ff.; Fox, Intermediate Economic Statistics, pp. 237-238, 278-282.

[23] Johnston, Econometric Methods, p. 6.

estimate but the value of the slope of the regression line is not affected. All three types of error are present in the samples under consideration. The errors in measurement in the dependent variable have been discussed above; among the omitted variables is the number of days the family left the town residence during the year and particularly during summer; finally the unpredictability of human response must be great with respect to a commodity which is relatively cheap and generally speaking taken for granted.

The variables in the equations are defined in the note to Table 21.

TABLE 21. MEANS AND STANDARD
DEVIATIONS -- SAMPLE I

Variable	Mean	Standard Deviation
V	198.539	66.747
L	62.443	53.926
Np	3.896	1.657
P	42.030	2.691
A	55.236	14.927
F	340.979	98.624
WUw	129.284	59.549
WUs	155.686	77.867
WUa	143.230	65.950

NOTES: WUa is the water use in gallons/day/dwelling unit (annual average)

WUs is the water use in gallons/day/dwelling unit (summer period average)

WUw is the water use in gallons/day/dwelling unit (winter period average)

V is the assessed sales value of residence in hundreds of dollars

L is the lot size in hundreds of square feet

Li is the area not covered by buildings in hundreds of square feet

Np is the number of persons in the dwelling unit

P is the variable price of residential water in cents/1000 gallons

Nb is the number of billing periods

A is the number of gallons/day/dwelling unit that are allowed with the minimum bill

F is the fixed bill for one billing period in cents

U is an error from the equation having the usual stochastic characteristics

Sample I

The means and standard deviations of all variables for this
sample are given in Table 21. The close clustering of price
values around the mean is indicated by the low standard deviation;
this lack of variation in price results in a large standard error
for the regression coefficient for this variable. In other respects
this sample brings out features that are repeated in the other
samples. The most important of these features is the lack of a
high 'peak' during the summer; the average water use in summer
is only 20.5 per cent higher than the winter average water use.
The 'peak' effect is partly averaged out by including a number
of days in the billing period that do not belong to the summer and
winter periods. Thus the average summer water use is lowered
by the inclusion of some days in autumn and spring while the
winter average is correspondingly increased.

The small 'peak' effect is also partly due to the fact that some
households spend part of the summer away from home; about 25
per cent of the households in the sample have an average summer
water use which is lower than that for the preceding or the
following period. The fitted equation would give a higher level
of 'explanation' if these items in the sample were screened out,
but it is preferable to keep these observations because the
sample is more representative of the universe of households in
the area.

The fitted equations for this sample are as follows:

$$WUa = 65.4 + 0.36V + 0.13L + 23.67Np - 0.32P - 0.23F \qquad (13)$$
$$ (0.12) \quad (0.15) \quad (3.80) \quad (2.65) \quad (0.09)$$

t-value $\qquad$ 2.88** 0.82 $\qquad$ 6.19** $\quad$ -0.12 $\quad$ -2.55*

$R = 0.70**$ $\quad S_Y = 65.94$

$R^2 = 0.49$ $\quad$ S.E. $= 49.19$ $\qquad$ F-value $= 11.53*$

$$WUa = 68.33 + 0.40V + 23.45Np - 0.61P - 0.20F \qquad (14)$$
$$ (0.11) \quad (3.8) \quad (2.62) \quad (0.08)$$

t-value $\qquad$ 3.75** $\quad$ 6.16** $\quad$ -0.23 $\quad$ -2.42*

$R = 0.69**$ $\quad S_Y = 65.9$

$R^2 = 0.48$ $\quad$ S.E. $= 49.06$ $\qquad$ F-value $= 14.3**$

* $\quad$ significant at 0.05 level

** significant at 0.01 level

The value of residence, the number of persons in the household, and the size of the fixed charge have significant regression
coefficients while size of lawn does not add enough 'explanation'.
This variable is fairly highly related to the value of residence
(Table 22). In theory the assessed value of the residence reflects
the lot size at least in part, and this is why lot size does not
contribute further to the explanation of the variance.

The same pattern is repeated with respect to the other two
dependent variables WUw and WUs; the equations with significant variables are:

$$WUs = 66.12 + 0.45V + 27.32Np - 0.59P - 0.23F \tag{15}$$
$$(0.13) \quad (4.59) \quad (3.16) \quad (0.10)$$

t-value $\quad 3.51** \quad 5.95** \quad -0.19 \quad -2.32*$

$R = 0.68** \quad S_Y = 77.87$
$R^2 = 0.46 \quad S.E. = 59.19 \quad F\text{-value} = 13.05*$

$$WUw = 140.54 + 0.24V + 22.00Np - 2.44P - 0.12F \tag{16}$$
$$(0.10) \quad (3.58) \quad (2.46) \quad (0.08)$$

t-value $\quad 2.33* \quad 6.15** \quad -0.99 \quad -1.46$

$R = 0.67** \quad S_Y = 59.9$
$R^2 = 0.44 \quad S.E. = 46.17 \quad F\text{-value} = 12.32*$

As one would expect the regression coefficients for assessed
sales value of residence (V) and number of persons in the household (Np) are greater for the summer average water use than for
the winter average water use; the coefficients for the annual
average water use lie between the coefficients for the other two
equations. The intensity of water use with respect to V and
Np increases during the summer periods.

Sample II
The means and standard deviations of all variables in this sample
are given in Table 23: data on the number of billing periods is
added to the list of variables shown in Table 21. Comparison of
the two tables shows that the means and the standard deviations
of all variables except price are very similar; this is what one
would expect when some items from the universe are picked at
random and added to other items from the same universe also

TABLE 22. SIMPLE CORRELATION COEFFICIENTS -- SAMPLE I

	V	L	Np	P	A	F	WUw	WUs	WUa
V	1.000								
L	0.636	1.000							
Np	0.192	0.126	1.000						
P	0.128	0.115	0.042	1.000					
A	0.356	0.378	0.204	-0.569	1.000				
F	0.518	0.527	0.263	0.487	0.441	1.000			
WUw	0.266	0.189	0.604	-0.144	0.196	0.051	1.000		
WUs	0.343	0.176	0.577	-0.092	0.137	0.045	0.845	1.000	
WUa	0.360	0.225	0.586	-0.096	0.145	0.049	0.930	0.956	1.000

NOTE: For explanation of symbols see note to Table 21.

TABLE 23. MEANS AND STANDARD
DEVIATIONS -- SAMPLE II

Variable	Mean	Standard Deviation
V	195.251	64.166
L	71.300	53.248
Np	3.989	1.690
P	45.450	14.318
Nb	4.527	0.886
A	52.197	18.164
F	316.980	121.480
WUw	127.764	61.434
WUs	158.038	81.317
WUa	142.919	67.913

NOTE: For explanation of symbols see note to Table 21.

picked at random. Similarly the regression coefficients with respect to V, Np obtained from sample I and sample II are quite similar; the coefficients from the larger sample have smaller standard errors.

The full equations for this sample are:

$$\text{WUa} = 178.9 + 0.31V + 0.19L + 22.08Np - 2.66P - 0.80\,Nb - 0.40A - 0.17F \quad (17)$$
$$\phantom{\text{WUa} = 178.9 + }(0.11)\quad(0.13)\quad\;(3.14)\quad\;(0.63)\quad\;(6.79)\quad\;(0.51)\quad(0.07)$$

t-value   2.8**   1.48   7.03**   -4.21** -0.12   -0.78   2.36*

$R = 0.73^{**}$ $S_Y = 67.9$
$R^2 = 0.54$ S.E. $= 48.2$ F-value $= 13.66^{**}$

$$\text{WUs} = 178.7 + 0.34V + 0.19L + 25.49Np - 3.34P + 6.82Nb - 0.5A - 0.17F \quad (18)$$
$$\phantom{\text{WUs} = 178.7 + }(0.13)\quad(0.15)\quad\;(3.81)\quad\;(0.77)\quad\;(8.24)\quad\;(0.62)\quad(0.09)$$

t-value   2.55**   1.25   6.7**   -4.3**   0.82   -0.82   -2.03*

$R = 0.72^{**}$ $S_Y = 81.32$
$R^2 = 0.52$ S.E. $= 58.54$ F-value $= 12.95^{**}$

$$\text{WUw} = 140.39 + 0.21V + 0.12L + 20.48Np - 1.78P - 4.29Nb + 0.08A - 0.15F \quad (19)$$
$$\phantom{\text{WUw} = 140.39 + }(0.10)\quad(0.12)\quad\;(3.03)\quad\;(0.61)\quad\;(6.55)\quad\;(0.49)\quad(0.07)$$

t-value   2.03**   0.98   6.67**   -2.93** -0.65   0.15   -2.2*

$R = 0.69^{**}$ $S_Y = 61.43$
$R^2 = 0.47$ S.E. $= 46.49$ F-value $= 10.58^{**}$

The variables L, Nb, and A have non-significant regression coefficients; L is partly reflected in the assessed sales value of residence. The amount of water allowed with the minimum bill A, has a high correlation with both F and P, i.e., when the minimum bill is high, more water is allowed with the minimum bill and when the variable price is high, the minimum bill tends to be low (Table 24). The number of billing periods (Nb) is not significantly related to any other variable; its non-significance may be due to its low variance (Table 23).

When the non-significant variables are omitted the fitted equations are as follows:

WUa = 131.26 + 0.37V + 22.15Np - 2.14P - 0.16F   (20)
           (0.10)   (3.07)   (0.48)   (0.06)

t-value   3.20**   7.22**   -4.5**   -2.56*

R = 0.72 S_Y = 67.9
R^2 = 0.52   S.E.   = 48.10   F-value = 23.36**

WUs = 152.28 + 0.41V + 26.15Np - 2.61P - 0.19F   (21)
           (0.12)   (3.74)   (0.58)   (0.08)

t-value   3.51**   6.98**   -4.5**   -2.4*

R = 0.71** S_Y = 81.32
R^2 = 0.50   S.E.   = 58.71   F-value = 21.66**

WUw = 115.72 + 0.26V + 20.43Np - 1.74P - 0.13F   (22)
           (0.09)   (2.93)   (0.45)   (0.06)

t-value   2.82**   6.9**   -3.8**   -2.07*

R = 0.68** S_Y = 61.43
R^2 = 0.46   S.E.   = 46.00   F-value = 18.6**

TABLE 24. SIMPLE CORRELATION COEFFICIENTS -- SAMPLE II

	V	L	Np	P	Nb	A	F	WUw	WUs	WUa
V	1.000									
L	0.511	1.000								
Np	0.133	0.173	1.000							
P	-0.290	0.091	0.021	1.000						
Nb	-0.086	0.279	0.153	0.402	1.000					
A	0.446	0.179	0.111	-0.765	-0.281	1.000				
F	0.543	0.251	0.135	-0.635	-0.331	0.725	1.000			
WUw	0.323	0.189	0.555	-0.308	-0.041	0.324	0.273	1.000		
WUs	0.374	0.236	0.538	-0.358	0.049	0.330	0.299	0.839	1.000	
WUa	0.396	0.245	0.549	-0.350	-0.008	0.329	0.302	0.936	0.957	1.000

NOTE: For explanation of symbols see note to Table 21.

105

Sample II(a)

The following equation is given to serve as a rough check of (20):

$$WUa = 114.04 + 0.34V + 21.08Np - 1.76P - 0.13F \qquad (23)$$
$$\qquad\qquad (0.09) \quad (2.86) \quad (0.36) \quad (0.06)$$

t-value $\qquad\qquad$ 3.88** $\quad$ 7.37** $\quad$ -4.95** $\quad$ -2.19*

$R = 0.72^{**}$ $S_Y = 66.50$

$R^2 = 0.52$ $\quad$ S.E. $= 46.95$ $\qquad$ F-value $= 26.4^{**}$

Examining the Residuals

In performing the regression analysis, a number of assumptions have been made about the errors from the equations viz., the errors (u_i) have zero mean and constant variance (σ^2) and are independently distributed with respect to the X-variables.

The residuals from equation (20), i.e., daily water use averaged over a year when $N = 91$, will be examined for expository purposes but the comments apply to all equations. The mean of the errors is 0.004 and 68 out of 91 items are within one standard error from this mean. The distribution of the 23 values of u_i that are greater or smaller than one standard error calls for comments because it throws light on residential water use patterns and has some bearing on subsequent discussion. Table 25 shows that the high residuals are positive (i.e., the distribution of residuals shows positive skewness).

The high positive residuals may indicate a water use-price function which is concave from above since the high positive residuals do not occur in the two municipalities with high price.[24] This type of curvilinearity has already been suggested from a priori considerations. The evidence is also consonant with a set of observations that have a lower limit but no upper limit.[25]

---

[24] The residuals from the equation (23) were also examined and confirm the distribution in Table 25 above; in other words the absence of high positive residuals in Newmarket is supported by the observations from Pickering Village.

[25] This high range in residential water use and positive skewness is reported by other investigators, e.g., T.R. Lee, Residential Water Demand and Economic Development, University of Toronto Department of Geography Research Publications No. 2 (Toronto: University of Toronto Press, 1969), p. 69, figure 5.

TABLE 25. DISTRIBUTION OF RESIDUALS

	Number of Observations	
	Positive	Negative
2 S. E. $> u_i > 1$ S.E.	5	11
$u_i > 2$ S.E.	7	-

The expected water use may be exceeded by any amount but the negative residuals cannot exceed the expected value itself.

There is some reason to expect the high (positive) residuals will occur when the price is lower; as price increases the large positive residuals will become less likely since careless uses of water will be cut down at high prices. Therefore, at high prices the variance will be lower and the assumption of homoscedasticity with respect to price is violated. Table 26 suggests that this is so in the present instance.

Acton points out that no robust test for homoscedasticity is known; all are sensitive to departures from normality in the residuals.[26] Both Hartley's test (which compares S^2max with S^2min) and Bartlett's test (which compares the weighted sum of the common logarithms of the individual group variances with the weighted logarithm of the mean of the group variances) were applied to the residuals from each of the price levels.[27] Both result in the rejection of the null hypothesis that the variance is homogeneous with respect to price, at 0.05 and 0.01 levels of significance.

Since the residuals are not normally distributed the tests may be positive because of this feature rather than lack of homogeneity. In the present case the interpretation is made more difficult by the fact that the observations from the price level with the lowest S^2 (Newmarket) are clustered around the mean for other variables

[26] F.S. Acton, Analysis of Straight-Line Data (New York: Wiley, 1959), pp. 89-91.

[27] Ibid., p. 98 ff. Also G.W. Snedecor and W.G. Cochran, Statistical Methods, 6th ed. (Ames: Iowa State University Press, 1967), p. 296; E.S. Pearson and H.O. Hartley, eds., Biometrika Tables for Statisticians, Vol. 1 (Cambridge: Cambridge University Press, 1966), p. 63.

TABLE 26. VARIANCE OF RESIDUALS AROUND REGRESSION

Price (c/1000 gallons)	30	39	44	45	71[a]	80
Variance	3891	2948	1469	2718	1120[a]	500

[a]These two values refer to Pickering Village and are obtained from equation (23) for illustrative purposes.

too (see Table 20). In fact when the Bartlett and the Hartley test are applied to the 102 observations less the 12 from Newmarket the hypothesis of homogeneity of variance in the residuals is not rejected at the 0.01 level of significance. Furthermore Table 26 does not show a clear systematic pattern of changes in variance of u_i with price.

It is tentatively concluded that water use data is expected to have a distribution of residuals which is positively skewed as well as heteroscedastic. Therefore, apart from any a priori expectations regarding the form of the function, a transformation of the data seems to be in order, if the standard errors of the regression coefficients are of primary importance in the investigation.

The regression coefficients obtained from the linear equations are still unbiased but would not have minimum variance. In other words the limits which are set on b_j are not reliable. In order to obtain the unbiased and best estimators, they must have minimum variance.[28] In order to satisfy the assumptions discussed above, a transformation of the data is required.

[28]Acton, Analysis of Straight-Line Data, p. 90; N. R. Draper and H. Smith, Applied Regression Analysis (London: Wiley, 1966), p. 80; P. G. Hoel, Introduction to Mathematical Statistics, 3rd ed. (New York: Wiley, 1963), pp. 228-233.

The Non-Linear Equations

Several curves may describe a set of data more or less equally well. In the present instance a curve is required which is concave from above with respect to the number of persons in the household (Np) and the assessed sales value of residence (V). In both respects there is no point of inflexion; therefore polynomial curves are ruled out. The exponential curves are also excluded since these would give a curve convex from above with respect to price and concave with respect to Np and V. The log Y and $\sqrt{Y}$ transformations are excluded for the same reason.

The square root transformation for both sides of the equation is acceptable on a priori grounds but the residuals show the same features as the linear function. The $\log_{10}$ transformation satisfies the assumptions of homogeneity of variance and normality of distribution of the residuals.

In order to satisfy the hypothetical relationship specified above the value of $\hat{b}$ for log Np and log V have to be less than one i.e., the exponent of Np and V has to be less than one; the slope of P and F will always be concave from above irrespective of the value of $\hat{b}$ for log P and log F provided the sign is negative.

In a previous section it was argued that the elasticity of demand with respect to marginal price should be greater for the summer water use than for the winter water use because the summer water demand includes some 'less essential' uses. Therefore, $\hat{b}$ should be expected to be higher in the summer equation. However, in order to obtain more accurate results on this particular problem information is required regarding the number of days the household resides in the city home.

The full equations for Sample II are set out in Table 27 (the term 'log' preceding the variable name signifies that logarithms to base 10 are taken).

The variables log L, log Nb, and log A are not significant and log A has the 'wrong' sign. In the present instance low log P are associated with high log F and high log A (Table 29). The variable log A is excluded from the set of equations of 'best fit'; it should be noted that the simple correlation of this variable with water use are positive — as is expected — and significantly different from zero.

TABLE 27.   LOG-LINEAR EQUATIONS

log WUa = 4.08 + 0.65 log V + 0.01 log L + 0.65 log Np - 1.29 log P - 0.46 log Nb - 0.33 log A - 0.35 log F   (24)

| | (0.15) | (0.09) | (0.08) | (0.30) | (0.29) | (0.20) | (0.17) |

t-value 4.38^{**} 0.06 7.67^{**} -4.34^{**} -1.58 -1.66 -2.07^{**}

$R = 0.77^{**}$ $S_Y = 0.22$

$R^2 = 0.59$ S.E. $= 0.15$ F-value $= 17.03^{**}$

log WUs = 4.18 + 0.62 log V - 0.03 log L + 0.67 log Np - 1.39 log P - 0.21 log Nb - 0.35 log A - 0.30 log F   (25)

| | (0.16) | (0.10) | (0.09) | (0.32) | (0.31) | (0.21) | (0.18) |

t-value 3.89^{**} -0.32 7.46^{**} -4.35^{**} -0.06 -1.69 -1.69

$R = 0.76^{**}$ $S_Y = 0.23$

$R^2 = 0.58$ S.E. $= 0.16$ F-value $= 16.04^{**}$

log WUw = 3.29 + 0.53 log V + 0.01 log L + 0.66 log Np - 0.92 log P - 0.46 log Nb - 0.06 log A - 0.37 log F   (26)

| | (0.17) | (0.98) | (0.10) | (0.34) | (0.33) | (0.22) | (0.19) |

t-value 3.15^{**} 0.01 6.90^{**} -2.7^{**} -1.4 -0.27 -1.97^{*}

$R = 0.71^{**}$ $S_Y = 0.22$

$R^2 = 0.50$ S.E. $= 0.16$ F-value $= 12.00^{**}$

TABLE 28.   MEANS AND STANDARD DEVIATIONS OF THE VARIABLES IN LOGARITHMIC FORM -- SAMPLE II

Variable	Mean	Standard Deviation
log V	2.268	0.141
log L	1.767	0.262
log Np	0.560	0.193
log P	1.640	0.116
log Nb	0.648	0.078
log A	1.683	0.185
log F	2.461	0.204
log WUw	2.054	0.223
log WUs	2.142	0.229
log WUa	2.105	0.218

NOTE: For explanation of symbols see note to Table 21.

The equations, after excluding log L, log Nb, log A are as follows:

$$\log WUa = 2.78 + 0.56\log V + 0.59\log Np - 0.93\log P - 0.31\log F \qquad (27)$$
$$(0.13) \qquad (0.08) \qquad (0.22) \qquad (0.14)$$

t-value   4.43**   7.26**   -4.14**   -2.26**

$R = 0.75$** $S_Y = 0.22$
$R^2 = 0.56$ S.E. $= 0.15$ F-value $= 27.5$**

$$\log WUs = 3.24 + 0.51\log V + 0.63\log Np - 1.07\log P - 0.35\log F \qquad (28)$$
$$(0.14) \qquad (0.09) \qquad (0.24) \qquad (0.15)$$

t-value   3.80**   7.29**   -4.50**   -2.40**

$R = 0.74$** $S_Y = 0.30$
$R^2 = 0.55$ S.E. $= 0.16$ F-value $= 26.45$**

$$\log WUw = 2.45 + 0.48\log V + 0.62\log Np - 0.75\log P - 0.24\log F \qquad (29)$$
$$(0.14) \qquad (0.09) \qquad (0.25) \qquad (0.15)$$

t-value   3.37**   6.82**   -3.03**   -1.56

$R = 0.70$** $S_Y = 0.22$
$R^2 = 0.49$ S.E. $= 0.16$ F-value $= 20.38$**

The sign and value of the regression coefficients are as expected. The residuals from equation (27) were examined for departures from normality in distribution and lack of homogeneity of variance. The average u is 0.0004 and 3 (mean-median)/S is equal to 0.0047. The S^2max/S^2min test gives a value of 2.712 which is less than the rejection level of 3.8 at the 0.05 level of significance. The residuals may be accepted as normally distributed and homoscedastic.

The equations (27) - (29) have two other important characteristics. The curves approach both axes asymptotically. This feature is due to the fact that zero cannot be represented on a logarithmic scale; for example as price approaches zero the curve becomes almost parallel to the Y-axis. On the other hand, as price increases there is no point at which demand becomes zero. It is, therefore, important not to extrapolate the equations beyond the range of the observations at hand.

TABLE 29. SIMPLE CORRELATION COEFFICIENTS OF
THE VARIABLES IN LOGARITHMIC FORM -- SAMPLE II

	log V	log L	log Np	log P	log Nb	log A	log B	log WUw	log WUs	log WUa
log V	1.000									
log L	0.495	1.000								
log Np	0.152	0.126	1.000							
log P	-0.318	0.068	0.003	1.000						
log Nb	-0.075	0.403	0.151	0.268	1.000					
log A	0.460	0.139	0.101	-0.833	-0.367	1.000				
log F	0.475	0.020	0.063	-0.798	-0.470	0.834	1.000			
log WUw	0.403	0.130	0.565	-0.313	-0.044	0.337	0.272	1.000		
log WUs	0.422	0.105	0.560	-0.394	0.021	0.332	0.306	0.853	1.000	
log WUa	0.464	0.131	0.561	-0.378	-0.030	0.342	0.311	0.936	0.964	1.000

NOTE: For explanation of symbols see note to Table 21.

Secondly, these equations show a constant price elasticity
of demand. It is attractive to hypothesize an increasing price
elasticity of demand as price increases and it may be possible
to find another form of the function that has the same desirable
statistical characteristics as the logarithmic form except for
its constant elasticity (e.g., taking the nth root). This curve
would then depart from linearity to about the same extent as the
logarithmic form of the function so that the elasticities along
the two curves would be similar. In addition, the multiplicative
form of the equation facilitates the comparison of the price
and income elasticities among the various localities or different
situations in the same locality (e.g., metered vs. unmetered
water use).

Price is the most important policy variable and it is worth -
while to examine the impact of price in more detail. Figure 12
shows the demand curve for residential water within the observed
price range when the family size (Np) is set equal to 4, the
assessed sales value of residence is $20,000 and the minimum
bill is $3.00. The demand curve in Figure 12 may be compared
with the hypothesized price-quantity relationship in Figure 11(b).

It was stated above that the demand for water is a composite

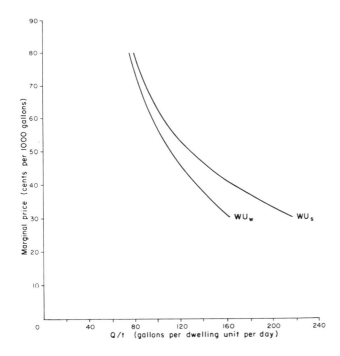

Figure 12. Demand for Residential Water in the
Toronto Region During Summer (WU$_s$) and During
Winter (WU$_w$)

demand and in Figure 11(b) it is hypothesized (a) that the slope
for $\sum_1^2 d_i$ is steeper than the slope for $\sum_1^3 d_i$ when the price is
low, (b) the demand curves for $\sum_1^2 d_i$ and for $\sum_1^3 d_i$ converge
and meet.

In Figure 12 the curve for the summer use (WUs), which
corresponds to the demand curve for $\sum_1^3 d_i$ in Figure 11(b),
is less steep than the demand curve for winter use (WUw). WUs
includes one or more added extra items of demand for water
(e.g., lawn watering, more frequent showers and so on). The
demand curves for WUs and for WUw in Figure 12 converge in
the manner suggested in Figure 11(b).

113

In terms of management possibilities: the reduction of water use for all uses combined for a given (small) price increase is greater when price is low than when price is high. At a low price 'wasteful' water uses are very responsive to price increases but at a high price less desirable water uses have already been discarded.

NON-METERED DWELLING UNITS:
GROUPS OF TOWNHOUSES

The Sampling Design

Sampling objectives and procedures
The purpose of the sampling design is to select about 40 observations of groups of dwelling units where the customers do not pay directly for water used. In this way one can obtain the average water use per dwelling unit in the group of townhouses and compare it with the water use of households who pay a variable price for water.

Generally speaking townhouses are metered as a group; they are the closest substitute to single-unit dwellings whose water use is not metered. The difference between the expected water use of a group of townhouses and of a group of metered single-unit dwellings may be ascribed in part to metering and price.

The observations were obtained from the municipalities closest to Toronto. The city of Toronto was not included because residential water use is not metered; in the municipalities of York and East York no observations of groups of townhouses for 1967 were available. The number of observations and the choice of municipalities were chosen arbitrarily on grounds of convenience (Table 30).

Preliminary inquiry did not turn up any meaningful basis for classification or stratification. Table 31 shows that the observations for groups of townhouses are distinguished by lack of variety as regards the average assessed sales value, the size of lot, and the size of the family.

TABLE 30. NUMBER AND LOCALITY OF
GROUPS OF TOWNHOUSES IN SAMPLE

Municipality	Number of Observations
North York	14
Scarborough	8
Etobicoke	9
Pickering Township	7
TOTAL	38

A list of groups of townhouses already rented in January 1967
was obtained from the Planning Department of the municipality.
Since the number of townhouses in 1967 was limited, the whole
list was used. For some observations the complete data from
the Water Accounts and the Assessment offices were not available
and these were screened out, leaving 38 items in the sample.

The Estimating Equations
The data in this sample may be used for regression purposes in
three forms; in the original form each observation is the total of
the number of dwellings included in the group. If there are k

TABLE 31. MEANS AND STANDARD DEVIATIONS FOR
AVERAGE VALUES OF 38 GROUPS OF TOWNHOUSES

| Variable | Natural Form | | Logarithmic Form | |
	Means	Standard Deviation	Means	Standard Deviation
V	165.79	24.04	2.21	0.065
L	25.56	5.30	1.39	0.17
Np	4.68	0.74	0.66	0.07
WUw	169.68	49.99	2.16	0.105
WUs	183.95	42.06	2.25	0.099
WUa	179.19	38.50	2.24	0.089

NOTE: For explanation of symbols see note to Table 21.

groups the relationship may be formulated as:

$$Y_k = f(X_{1k}, \ldots X_{mk})$$ (30a)

where there are m variables. In this form the relationship would not add much to the understanding of the patterns of water use. All the variables are some function of the number of dwelling units that go to make up an observation (n_1). The elements of the simple correlation matrix for this form of the equation are all greater than 0.93.

It is valid to state the relationship in terms of the average increase in the dependent variable with the average increase in the independent variables. If $Y_k \big/ n_k = \overline{Y}_k$ then

$$Y_k = f(X_{1k}, X_{2k}, \ldots X_{nk})$$ (30b)

This form of the hypothetical relationship gives unbiased estimates of the regression coefficients but they would not have minimum variance.[29]

The logarithmic transformation is used for the reasons given in the discussion of water use in single-unit dwellings. This transformation also yields easy comparison of the elasticities of water use with respect to the number of persons in the household and the assessed sales value of the residence in the two types of residential units.

The transformation to the logarithmic scale is justified only if the townhouses making up the kth group do not differ very

---

[29]The variance of the mean in the kth group e.g., $S_{\overline{Y}_k}^2$ is given by

$$S_{\overline{Y}_k}^2 = S_u^2 / n_k$$

if the values of the averages of groups of townhouses in repeated samples are assumed to be normally distributed around the regression plane. Therefore the means of the groups of dwelling units will have a variance from the regression plane inversely proportional to n_k or a standard deviation proportional to $1/\sqrt{n_k}$. Correct standard deviations of the mean values from the regression equation are abtained by weighting the observations by $\sqrt{n_k}$:

$$\sqrt{n_k} \cdot Y_k = f(\sqrt{n_k} \cdot X_{1k}, \ldots \sqrt{n_k} \cdot X_{mk})$$ (30c)

The detailed application of weighted least squares to this problem is set out by Howe and Linaweaver, "The Impact of Price on Residential Water Demand," pp. 31-32. See also Johnston, Econometric Methods, pp. 207-211.

much among themselves so that the value of $\log \bar{Y}_k$ (where $\bar{Y}_k =$

$$\sum_{i=1}^{n_k} Y_{ik}/n_k) \text{ is not significantly different from } \sum_{i=1}^{n_k} \log Y_{ik}/n_k.$$

This assumption is a reasonable one to make because a block of townhouses consists of dwelling units built with the needs of households sharing the same style of life; families living in a block of townhouses may be expected to be relatively homogeneous with respect to income level and size of household. The transformation into logarithms of average values (i.e., $\bar{Y}_k$) introduces a bias in the results to the extent that the dwelling units in each group of townhouses vary from the mean assessed sales value of residence and mean household size for the group. The correction for this bias requires further investigation and more complete data about the townhouse units.

Table 31 gives the means and standard deviation of the group averages. The single-unit dwellings had only a modest 'peak' from average winter use to average summer use (24 per cent). Lack of information about the number of days the family stayed away from the residence (e.g., holidays) accounts for the small increase at least in part. The townhouses have a lower 'peak' still (8 per cent) as may be expected from the smaller lawn size for the townhouses. This means that the summer water use to be expected in unmetered single-unit dwellings is conservatively estimated by the water use in townhouses; the effects of metering and marginal pricing are probably greater than the present results indicate. Even so, the water use in townhouses as shown in Table 31 is considerably higher than the water use in metered houses shown in Table 23 particularly during the winter period when the comparison is more valid.

The equations for the unweighted observations are:

$$\log WUa = 0.17 + 0.72\log V + 0.14\log L + 0.43\log Np \qquad (31a)$$
$$(0.20) \qquad (0.07) \qquad (0.19)$$

t-value $\qquad\qquad$ 3.5** $\qquad$ 1.88 $\qquad$ 2.21*

$R = 0.58**$ $\quad S_Y = 0.09$

$R^2 = 0.34$ $\quad$ S.E. $= 0.08$ $\qquad$ F-value $= 5.86**$

log WUa = 0.34 + 0.71logV + 0.49logNp   (31b)
               (0.21)   (0.20)

t-value   3.38**   2.45*

R = 0.52** S_Y = 0.09
R^2 = 0.27   S.E.   = 0.08   F-value = 6.54**

log WUs = 0.30 + 0.70logV + 0.62logNp   (32)
               (0.24)   (0.22)

t-value   2.9**   2.76**

R = 0.52** S_Y = 0.09
R^2 = 0.27   S.E. = 0.78   F-value = 6.55**

log WUw = 0.12 + 0.81logV + 0.44logNp   (33)
               (0.26)   (0.24)

t-value   3.16**   1.8

R = 0.48** S_Y = 0.11
R^2 = 0.23   S.E. = 0.10   F-value = 5.8**

The main objective of this least-squares analysis is to obtain accurate regression coefficients but the low value of R calls for some comment. Tables 23, 29, and 31 show that the observations for groups of townhouses are characterized by low standard deviations including those for the dependent variables. The observations are concentrated along part of the 'regression line' and this results in a low R.[30]

One should expect the values of the averages of the groups of townhouses to cluster closer to the regression plane than the values of the individual single-unit dwellings. In fact the standard error of the estimate for the average townhouse is 0.078 compared with 0.147 for the single-unit dwellings.

The coefficients of winter water use in single unit dwellings and groups of townhouses with respect to assessed sales value of residence (V) and number of persons in residence (N_p) are

---

[30] This is easily shown by considering the formula $R^2 = 1 - s^2_{YX}/s^2_Y$. If the variance s^2_{YX} is assumed to be constant. R^2 becomes larger as the variance of Y gets larger.

TABLE 32. EXPONENTS OF V AND N_p FOR RESIDENTIAL WATER
USE DURING WINTER IN THE TORONTO AREA

	Single-Unit Dwellings	Groups of Townhouses
V	0.48 (0.14)	0.81 (0.26)
N_p	0.62 (0.09)	0.44 (0.24)
SOURCE:	Equation 29	Equation 33

summarized in Table 32. The water use during winter should
be comparable in the two types of housing; during summer the
smaller lot size of townhouses makes comparison more difficult.

The coefficients or elasticities for N_p do not seem to be
substantially different for the two types of housing. The coefficients for V suggest that given the number of persons in the
household, water use is more responsive to differences in income
level for families living in townhouses. It may be that these
families have more water-consuming household appliances
particularly since townhouses are rented with a wide range of
semi-durables; since water is 'free' their use of such appliances
would be more frequent and with less thought to water conserving
habits.

THE COMBINED SAMPLE OF SINGLE-UNIT
DWELLINGS AND TOWNHOUSES

The group averages from the 38 observations on townhouses
are combined with the 91 observations of single-unit dwellings.
There is an aggregation problem since averages are combined
with observations on individual consumers. Weighting the
observations by $\sqrt{n_k}$ reduces this problem.

The linear function is used in this instance since in log-linear
demand functions, the value Y (water use) approaches infinity

119

as the value X (price) approaches zero. However, even at a price of zero the amount of water used is finite.[31]

It is possible to test whether the coefficients for townhouse observations and single-unit observations are significantly different. The equation is set out as follows:

$$WUw = a + b_1 V + b_2 L + b_3 N_p - b_4 p + b_5 X_5 + b_6 X_6 + b_7 X_7 \qquad (34a)$$

where V, L, N_p and p have the same values as in the equations set out so far and where:

p = 0 for unmetered observations;

p > 0 for metered observations;

X_5 = 1 for unmetered observations

X_5 = 0 for metered observations;

X_6 = V. X_5 = V for unmetered observations;

X_6 = V. X_5 = 0 for metered observations;

X_7 = N_p. X_5 = N for unmetered observations;

X_7 = N_p. X_5 = 0 for metered observations.

For metered observations the equation is:

$$WU_w = a + b_1 V + b_2 L + b_3 Np - b_4 p \qquad (34b)$$

and for unmetered observations the equation is:

$$WU_w = (a + b_5) + (b_1 + b_6) V + b_2 L + (b_3 + b_7) Np \qquad (34c)$$

---

[31] Another possible specification of the relationship that would retain the property of constant elasticity with respect to N_p, V and P but have a finite value for the dependent variable when price = 0 is the following:

$$WU = \frac{A X_1^b \dots X_n^b}{(p + c)^b}$$

where C > 0 so that WU approaches zero when (p + c) = 0 but WU is positive and finite when p = 0. The effect is to reduce the slope of the regression line with respect to price when price is low.

In the present instance the inclusion of a shift variable affects the price coefficient substantially even when C is made equal to 70 and the coefficient of the shift variable is small and non-significant. More observations on price (particularly below 30 cents/ 1,000 gallons) would indicate the nature of the slope of the price line in the lower range.

The fitted equation from the weighted observation is:

$$WUw = 77.9 + 0.12V + 0.11L + 19.2Np - 1.25p - 191.0X_5 + 0.76X_6 + 10.24X_7 \quad (35)$$

$$(0.21) \quad (0.23) \quad (6.3) \quad (0.8) \quad (79.0) \quad (0.26) \quad (8.74)$$

$R = 0.97$ $\quad S_Y = 427.0$

$\quad\quad\quad$ S.E. $= 100.0$ $\quad$ F-value $= 278.6$

The coefficients of lot size (L) and X_7 are not significantly different from zero but they have the expected sign. The magnitude of the coefficients of V, Np and p are all within one standard deviation of those for single unit dwellings (see equation [22] above). However, it is clear that the inclusion of the shift variable (X_5) and X_6 and X_7 (the interaction effect) affect the variance of these variables.

When L and X_7 are omitted the fitted equation is:

$$WUw = 54.4 + 0.15V + 24.9Np - 1.2p - 132.0X_5 + 0.65X_6 \quad (36)$$

$$(0.17) \quad (4.30) \quad (0.77) \quad (64.0) \quad (0.24)$$

$R = 0.97$ $\quad S_Y = 427.0$

$\quad\quad\quad$ S.E. $= 100.0$ $\quad$ F-value $= 423.9$

When X_5 and X_6 are omitted the result is:

$$WUw = -4.75 + 0.48V + 20.4Np - 0.94p \quad (37)$$

$$(0.11) \quad (4.1) \quad (0.24)$$

$R = 0.97$ $\quad S_Y = 427.0$

$\quad\quad\quad$ S.E. $= 102.0$ $\quad$ F-value $= 534.0$

The conclusions from the combined sample can only be tentative. It seems that townhouse observations behave markedly differently with respect to V (i.e., they use more water per value of residence), possibly because they have more water-using appliances which are used with less care since water is 'free'. It also seems that the coefficient of price is reduced when observations with zero price are added to single-unit houses. One may speculate that at low prices satiation is approached; the use of a couple of thousand gallons per billing period would cost very little more (e.g., 60 cents when the commodity charge is 30 cents/1,000 gallons). However this hypothesis can only be tested properly in a situation where more observations on the lower values of price ($p < 30$) are available, preferably for single-unit dwellings.

Conclusion

The theoretical relationships set out at the beginning of this chapter were derived from simple assumptions about the manner and reasons for residential water use. Not all the hypotheses set out have been validated because of lack of variance and/or because of inter-relationships among some of the independent variables.

Two important tentative results may be singled out: (1) the functional form of the equation is curvilinear in order to reflect the composite demand for water in and around the house; (2) the efficacy of pricing as a policy variable has been established and the ranges over which price increases are more effective have been indicated.

The results of the theoretical and empirical work presented in this chapter are discussed further and applied in the next chapter on policy alternatives. The appendices to this chapter contain some additional evidence about the fit of the equations, the errors in the observations of summer and winter water use and the effects of metering and pricing.

APPENDIX TO CHAPTER III — A

Some Additional Evidence

The fitted equation of water use in single-unit metered residences was applied to some observations which were not used in the fitting of the regression equation. It is possible to compare the expected value as computed from the fitted equations (27), (28), (29), with some observed values of water use; since the main interest of this part of the study is to observe how the level of water use in a community (or subdivision or municipality) would respond to price changes, it is worthwhile to make the comparison between the computed water use and the observed average water use of a group of similarly assessed residences.

In Table 33 the comparison is made for five items. The winter and summer use for Pickering Village is not available. The close correspondence of the observed average and the explained value computed from the equation is encouraging and lends confidence to the use of the fitted equations in the Toronto Region. The only exception is item 2, viz., a group of residences in Etobicoke whose income level is beyond the range of observations that went into the sample. The warning against extending the use of the fitted model beyond the range of the observed value bears repitition.

TABLE 33. WATER USE IN FIVE GROUPS OF SINGLE-UNIT METERED DWELLINGS

Item	Number of Dwellings	Values of Variables				Average Observed	Water Use (Gallons/Day/Dwelling Unit)				
		P	F	V	Np		Annual Computed	Average Obs.	Summer Comp.	Average Obs.	Winter Comp.
1 Etobicoke	7	39	270	17165	3.143	127	127	140	131	101	101
2 Etobicoke	4	39	270	47090	3.75	299	249	362	264	182	212
3 Pickering V.	6	71.25	285	16770	3.66	96	---	---	---	---	---
4 Pickering V.	4	71.25	285	25500	3.53	106	103	---	---	---	---
5 Newmarket	6	80	100	21240	4.0	114	116	114	119	105.5	104

NOTE: P   is the variable price of residential water in cents/1000 gallons.
        F   is the fixed bill for one billing period in cents.
        V   is the assessed sales value of residence in hundreds of dollars.
        Np is the number of persons in the dwelling unit.

The municipality of Mimico became part of the borough of Etobicoke in 1967. The Mimico residences are on a flat-rate for water and their water consumption is not metered. In 1968 the Etobicoke Department of Engineering set up a study to compare the water consumption by and the revenue from metered and unmetered residences with a view to installing meters in the residences formerly in Mimico. [1]

Three groups of residences were selected:

1. 16 houses on North Carson Street which have been individually metered for some time;
2. 16 houses on 11th Street, formerly not metered, now metered individually (ARB meters) but still billed on a flat-rate basis;
3. 47 houses on 10th Street, formerly unmetered and now connected to a master meter. This group served as a control group in case the households in group (2) reacted positively to the installation of meters.

All consumption was read monthly during 1968 and all houses were in the same assessment group ($5000 to $6000) and had an assessed sales value of about $19,000.

Reference was made above to 'errors in measurement' of the dependent variables. These errors are due to (a) the fact that individual billing periods vary in duration; (b) the summer and winter bills may include parts of spring or autumn thus lowering (increasing) the summer (winter) average. The data from the Etobicoke study sheds some light on probabilities that the inclusion of certain months contributed to the underestimation of the summer average water use or to an overestimate of the winter average water use.

Table 34 shows that for 32 observations the month with the least chance of understating the summer average is July. Even if one picked an observation for the months of June, July, and

---

[1] The data for this Appendix were made available for this study by Mr. A. Bernard, Chief Engineer, and Mr. D. Gilpin, Engineer, Department of Engineering, Borough of Etobicoke.

TABLE 34. ETOBICOKE STUDY: FREQUENCY FOR MONTHS WITH
WATER USE BELOW SUMMER AVERAGE (May through August, 1968)

Frequency out of 32 observations					
April	May	June	July	August	September
19	12	14	9	14	24

August, there is still a chance of 28 out of 32 that either June or
August is below average and a chance of 1 in 4 that both are below
average. Therefore, the water use average for summer reported
in this study is most likely a conservative estimate of the true
'summer' use when several extra water-complementary activities
are added to the usual water demand. Table 35 indicates that the
average water use for winter is probably overestimated and this
accounts for the low peaking effect noted above. It is to be
expected that if these errors of measurement of the dependent
variables were removed (e.g., by monthly reading of water
meters or daily readings), the difference in the slopes of the
demand functions in Figure 12 above would be more marked; in
other words the impact of price on peak water use would be even
more clear.

The average daily consumption per capita is given in Table 36.
(The observed water use is quite close to the expected water use
obtained from the fitted equations.) The level of water use in the
group of residences on 11th Street which are individually metered
but billed on a flat-rate basis is little different from the 'unmetered'
group; the other metered groups of residences show a reduction

TABLE 35. ETOBICOKE STUDY: FREQUENCY FOR MONTHS WITH WATER
USE GREATER THAN WINTER AVERAGE (December through February, 1968)

Frequency out of 32 observations						
January	February	March	April	October	November	December
8	11	16	14	11	13	8

TABLE 36. WATER USE AND COST OF WATER IN FIVE GROUPS
OF HOUSES IN ETOBICOKE IN 1968

Type	Street	Number of Homes	Number of Occupants	Average Daily Consumption (Gallons/Capita/Day)	Average ($) Paid per Residence	Average ($) Paid per 1000 Gallons
Unmetered	11th Street	16	49	56.2	25.32	40.0
Unmetered	10th Street	47	162	59.5	25.32	33.8
Metered	North Carson	16	82	38.4	40.36	56.0
Metered	Smithfield	16	62	38.2	32.04	59.3
Metered	Stephen	16	45	34.8	21.24	59.3
Metered	Merriday	15	79	38.8	42.44	56.6

of about 35 per cent from the unmetered consumption rate. It
is clear that the marginal price makes the difference rather
than the metering per se.

Table 36 makes the important points (a) the marginal price
of water is less than the average price for many metered
customers (the marginal price in 1968 in Etobicoke was 47
cents/1000 gallons); (b) unmetered consumers in Etobicoke paid
less for water than metered customers: the total bill is less
and the average price is lower too.

IV
Evaluating policy alternatives

EVALUATING POLICY ALTERNATIVES

Introduction

Municipal water systems take only a small fraction of total water withdrawals for all uses but the decisions to allocate public resources to municipal water supply merit attention for several reasons.

Since classical times the demand for urban water has required relatively heavy investment in engineering works designed to transport, purify, and distribute water of the desired quality to urban consumers. In addition the effluent from urban sewage systems requires extensive and expensive treatment unless it is allowed to degrade the quality of the receiving waters.[1]

Municipal water is the most valuable both in terms of investment required and also in terms of user's utility. Hirshleifer, DeHaven, and Milliman point out that the total value in exchange of water is

---

[1] The Ontario Water Resources Commission (OWRC) approved about $1 billion of water pollution control projects in the eleven years 1957-1967; this amount is about double the approved capital expenditure on water supply projects.

low in relation to the aggregate value in use, so that the importance of decisions relating to the allocation of resources in the municipal water field may be greater than is suggested by the small money value involved. [2]

Wollman has worked out the value added per acre-foot in the San Juan and Rio Grande basins in New Mexico; the figures do not take into account the cost of water to the user (Table 37). The value of water to the community in dollar terms is several times greater for industrial/municipal uses than for agricultural and recreational uses.

Renshaw has computed the value of an acre-foot of water for 1950, defining value as the amount paid for municipal, industrial, and irrigation water. As the author points out these figures are intended to be illustrative of possibilities for making comparisons rather than a final word as to the value of water in various uses. In fact they represent the average amount paid for an acre-foot of water by various categories of water users. Domestic consumers pay about $100.19, industrial users $40.73, and irrigation farmers $1.67. [3] The ratio of the value of municipal to irrigation water is of the same order of magnitude in the two studies. The high value in use and the high value in exchange of municipal water justifies scrutiny of the policy towards this use of water.

Policy decisions affecting the level of the demand should be examined for another reason; viz., meeting the demand for municipal water is given a high priority by policy makers. A report on the "Basic Principles of a National Water Policy" presented to the American Water Works Association states that "first priority should be given to providing water for people for use in their home and urban activities."[4]

Given the high utility for a supply of treated, piped water in the home, the high cost involved in providing the service and the high priority accorded to this service, it is difficult to account for lack of interest in the factors affecting demand for residential

---

[2] J. Hirshleifer, J.C. DeHaven, and J.W. Milliman, Water Supply: Economics, Technology and Policy (Chicago: University of Chicago Press, 1960), p. 88.

[3] E.F. Renshaw, "Value of an Acre-Foot of Water," Journal AWWA 50, no. 3 (1958):303-9.

[4] AWWA Committee Report, "Basic Principles of a National Water Resources Policy," Journal AWWA 49, no. 7 (1957):831.

TABLE 37. GROSS PRODUCT PER ACRE-FOOT (U.S. Dollars)

	San Juan Valley	Rio Grande Valley
Agriculture	27.90 to 28.60	44.40 to 51.00
Municipal/Industrial	1255.00 to 2810.00	3041.00 to 3989.00
Recreation	--	212.00 to 307.00

SOURCE: N. Wollman, The Value of Water in Alternative Uses
(Albuquerque: University of New Mexico Press, 1962), p. 31.

water that are relevant to demand management. The widespread disregard of price by the professionals in the municipal water supply industry (both in projecting future demand and in pricing the commodity) suggests that the 'shortage of water' is a technical absolute; in fact it is partly a function of price and the metering/ pricing of residential water is the most important policy variable at the disposal of management. [5] Before discussing the present pricing system and its shortcomings in detail, it is useful to examine briefly the cost structure of the industry.

THE COST STRUCTURE OF MUNICIPAL WATER SUPPLY SYSTEMS

The production of municipal water involves a heavy initial outlay and a relatively modest annual outlay for maintenance and operation. Under these conditions average costs would decline

[5]In developing countries one has to take into account the type of service that residents are willing and able to pay for at a given level of development (e.g., public standpipes, single-tap services). In a North American context the question is more marginal i.e., one has to decide between a little more water or a lot more water rather than between a water service and none. T.R. Lee and G.F. White, et al. in the works referred to previously have useful discussions on the management opportunities in developing countries while the marginal nature of the decision to allocate resources for water development projects is raised in J.M. Milliman, "Economic Aspects of Public Utility Construction," Journal AWWA 50, no. 7 (1958):839-845.

until capacity is reached. In Figure 13 the short-run average cost (SAC) declines within each step. [6]

The average cost of municipal water depends upon many factors including the ratio between fixed and variable costs, the efficiency of the operation, the scale of operations, the source of water and degree of treatment required, the distance from source to the point of consumption, the relative importance of 'big' consumers, and the amount of storage required to meet peak demands.

The average cost derived by dividing total expenditures by total consumption[7] gives only the 'book' average cost. This depends, in addition to the factors mentioned above, on the amount of depreciation and the rate of repayment decided upon by the management of the particular water supply system.

In Ontario in 1966 it cost about 3 cents per 1000 gallons for pumping treatment of water from a local well; nearby surface water (lake or river) cost from 1 to 10 cents (depending on water quality) and water pumped through regional pipelines was estimated to cost 10 to 35 cents. [8]

An Increasing Cost Industry

Some municipal water engineers who were interviewed were of the opinion that long-run average costs were horizontal for water supply systems of each source. For example, in Figure 13 as capacity is reached in one plant of lake or well water, another unit is brought into operation. In fact economies of scale would be effective for a long time period because of fixed overhead costs (e.g., the number of engineers and managers is not proportional to the total quantity of water produced).

[6]The influence of capacity utilization and size of plant upon average costs is generally accepted. For some empirical verification using historical adjusted plant investments, see Lawrence C. Hines, "The Long-Run Cost Function of Water Production for Selected Wisconsin Communities," Land Economics 45, no. 1 (1969), table I, pp. 136-137.

[7]Such information is available for Ontario for 1934 onward in Ontario Department of Municipal Affairs, Annual Report of Municipal Statistics, Waterworks Section (Toronto: Queen's Printer, annual).

[8]J. A. Vance, K. E. Symons, and D. A. McTavish, "The Diverse Effects of Water Pollution on the Economy: Domestic and Municipal Water Use," in Pollution and Our Environment, Background Paper A 4-1-5, Canadian Council of Resource Ministers Conference, Montreal, 1966. Vol. 1. (Ottawa: Queen's Printer, 1967), p. 11, figure 2 and p. 14.

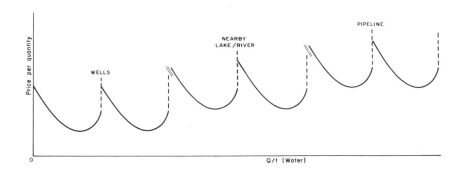

Figure 13. Hypothetical Short-Run Average Cost Curves for Municipal Water Supply Systems in Southern Ontario

The municipal water supply industry in southern Ontario as a whole may be said to exhibit a general rising trend in average cost over time because as the supply of water from wells proves to be insufficient for growing municipalities, the production process is changed to higher-cost surface water (lakes or river) and eventually to pipeline water if this source proves to be insufficient.

This trend is shown by the studies made for the Lake Huron-London pipeline. The cost of production, distribution, and administration were reported to be 17 cents/1000 gallons for well water (1963 dollars); 20 cents for wells and river water; 27.7 cents (dropping to 22 cents by 1985) for Lake Huron and well water and 28.3 cents (dropping to 22.4 cents by 1985).[9]

The investment cost for source development is the main contributor to the increase in cost. Wells in the London area were to be developed at a cost of $257,700 per Million Gallons Daily installed; the Lake Huron water costs about $623,600/MGD

[9]Engineering Board of Review, City of London Report on Water Supply (London, Ontario, December 1962), p. 20 ff.

131

(i.e., more than double).[10] One MGD can serve about 8000 people (at 125 gpcd) and the increase in investment per capita is $45.7.

The discussion so far referred to municipal water supply for all uses. Cost data for residential water production are not available since municipalities do not keep separate records for water use categories. However, there are two basic reasons for expecting the costs of production of residential water to increase over time.

1. As new subdivisions are opened away from the Lake (or other sources of water), pumping and related costs should increase to overcome the friction of distance and this cost is additional to the increase in pumping costs per million gallons produced; mains transmission average costs would also tend to increase slightly if the density of building is reduced.

2. Peak demands for residential water are increasing faster than average daily demands. This is due to higher income and larger lawns. Supply facilities (source development, treatment, distribution, and storage facilities) are expanded to cope with the higher peak demand and there is more excess capacity during off-peak periods. The proportion of idle plant during the off-peak season will increase over time thus increasing average costs.

Wolff notes that peak daily and hourly consumption have shown a pronounced upward trend. In suburban areas around Baltimore the maximum day/average day ratio (expressed as a percentage ratio) was increasing by 1 to 4.5 points per annum; in the city the increase was 0.9 of one point per annum.[11] Table 38 shows that the annual rate of increase is low in large municipalities and high in smaller (and more heavily residential) municipalities.

Table 39 provides more direct evidence of the fact that the provision of storage facilities to cope with increasing peak demands is more than keeping pace with the growth in the

---

[10] James F. MacLaren (Associates), Report to the Public Utilities Commission of London on Waterworks Development to 1985, (London, Ontario: Public Utilities Commission, December 1961), pp. 17, 66.

[11] J.B. Wolff, "Forecasting Residential Water Requirements," Journal AWWA 49, no. 3 (1957):225-235.

TABLE 38. RELATION OF POPULATION SERVED TO INCREASE
PERCENTAGE RATIO OF PEAK TO AVERAGE DAILY CONSUMPTION

Population (000)	Annual Increase Percentage Ratio	Number of Cities	Range
Under 10	4.3	6	-4.0 to 10.0
10 - 25	2.1	9	-0.7 to 8.0
25 - 50	2.0	6	0.3 to 4.5
50 - 100	1.8	5	0.5 to 4.0
100 - 250	1.6	11	-0.7 to 2.7
250 - 500	1.4	6	0.8 to 4.3
500 - 1000	1.0	3	0.0 to 1.5
over 1000	1.0	7	0.9 to 1.2

SOURCE: J. B. Wolff, "Forecasting Residential Water Requirements,"
Journal AWWA 49, no. 3 (1957):232.

total water use. Data were available for 9 municipalities in
Metropolitan Toronto and region whose 1967 population was
15,000 or more. London, Ontario was added to this list. The
increase in storage facilities between 1957 and 1967 is impressive;

TABLE 39. STORAGE FACILITIES IN MUNICIPAL WATER SUPPLY
SYSTEMS IN METROPOLITAN TORONTO AND REGION, 1957 AND 1967

	Storage Facilities (Million Gallons)		Gallons of Storage/ Million Gallons Consumed	
	1957	1967	1957	1967
Metro Toronto	121.975	220.320	2164	2864
Barrie	1.200	2.695	1974	2836
Guelph	1.500	5.000	852	3226
Hamilton	N.A.	82.360	N.A.	3694
London	22.000	39.000	5083	6358
Oshawa	5.250	5.500	2511	1633
Brampton	1.250	7.250	3676	7323
Burlington	2.314	6.900	3788	3896
Oakville	0.525	4.250	800	3233
Richmond Hill	Nil	1.000	Nil	N.A.

SOURCE: Figures of total consumption and total storage from Ontario
Department of Municipal Affairs 1957 and 1967 Annual Report of
Municipal Statistics.

more relevant is the increase in all cases but one (Oshawa) of the amount of storage provided per million gallons of water used. Provision of storage facilities is increasing faster than water consumption.

It may be concluded that residential water is a commodity whose supply increases in discrete quantities, that within each unit of increment the average costs decline until capacity is reached and that the marginal costs of supplying residential water are increasing in the long run.[12]

In Figure 14 the curve $SRMC_2$ refers to the short-run marginal cost curve that applies to the increment unit marked II. $SRMC_2$ should be below SRAC when the latter is falling, equal to SRAC when it is at its minimum and then it rises steeply when design capacity is reached.[13]

In Figure 15 let increment unit III be the latest addition to the water supply system. If the demand curve is D_1, short-run marginal cost pricing will produce a loss; as demand shifts (i.e., increases) to D_2 the loss will be reduced and at D_2 total revenue will be equal to total cost under marginal cost pricing (since price = marginal cost = average cost). At higher demand than D_2 marginal cost will be higher than average cost.

At Q_0 where the long run marginal cost (LRMC) intersects SRMC both are equal to price under marginal cost pricing. This point shows the optimal size of plant and if demand grows beyond D_4, then an increase in plant is called for.

In water supply, increments to the plant have to be made in discrete, indivisible quantities in order to take advantage of economies of scale in construction; the new addition to the plant should be built when the increase in aggregate value in use (i.e., the increase in area under the demand curve) is equal to the increase in aggregate cost of building the new plant. In other words the increase in total utility is equal to the increase in total cost.

---

[12]One may also point out that pollution control costs increase as the rate of water use increases; as the quantity of effluent increases the level to which the receiving water has to be cleaned also increases. However, this is an external diseconomy passed on to pollution control systems; the increase in cost to water supply due to increased water pollution is negligible.

[13]The cost analysis in this section follows the more detailed discussion by Hirshleifer, DeHaven, and Milliman, Water Supply: Economics, Technology and Policy, pp. 94-98.

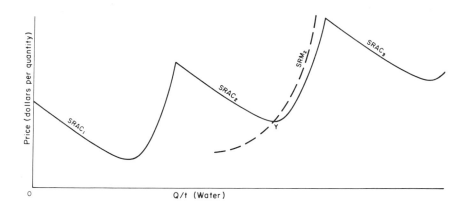

Figure 14. Short-Run Average and Marginal Cost Curves of
Municipal Water Supply

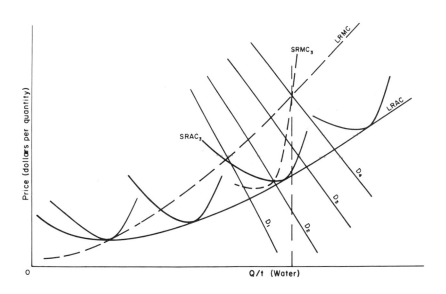

Figure 15. Average and Marginal Cost Curves of Municipal
Water Supply

135

PRICING RESIDENTIAL WATER

Some Guidelines

Municipal water is a commodity produced and sold by a monopoly which acts as a trustee of the public good. The monopoly has some leeway in setting prices and the price set depends on the objectives of those setting the price. There are seven objectives that may be distinguished in the price setting of municipal water:

1. To recover expenditures; this is the most widespread objective in the municipal water supply industry.
2. To leave a small margin of profit to the taxpayer so that further expansion may take place without undue difficulty.
3. To subsidize certain users (e.g., new industry) in the hope of gaining benefits from an expanded tax base.
4. To make a small margin of profit which is used to reduce the level of municipal taxation; water is one of the few municipal services that could be operated at a profit.
5. To redistribute income; this requires that small users pay a lower price than large users of water.
6. To allocate resources efficiently by setting price equal to marginal cost, thus equating marginal utility with marginal cost of production and with price.
7. As a tool in the hands of management; pricing can be directed to decreasing average and maximum day demand.

The first four aspects of pricing are the most commonly observed in North America while the last three aspects are conspicuous by their absence from the water schedules described in the literature.

The subject of water rates is controversial not only because the set objectives may differ, but also because the cost of water, the ratio of demand of the various user classes, the intensity of seasonal and daily demand and other factors vary from one water system to another. A rate structure of universal or even wide applicability is impractical. But the general approach to water rates can be set out on logical grounds so that the water manager who is free to follow such advice may implement a policy of pricing water that is calculated to reach the objectives he chooses.

The ideal water rate schedule should be attractive on grounds of both logic and applicability; it is possible to set rates so that

the first three of the following principles are met on a sine qua
non basis; the fourth principle would make the pricing system
more logical and humane but it involves imposing a political
value-judgement since it implies redistribution of income. [14]

The principles on which an ideal water rate schedule should
be based are proposed as follows:

1. It should be easy to administer and to understand.
2. Expenditures for the production and distribution of residential
water should be recovered through water revenue from residential users.
3. The price of the last gallons used should reflect the cost of
producing them; when the marginal cost is known this should be the
price of the last units of consumption.
4. If possible the first few gallons/capita/day (or per dwelling
unit per day) should be sold at a nominal price to reflect the
returns that accure to society from the use of clean water for
essential purposes.

Before discussing price as a policy instrument a short
description of the present practice and the economic solution to
rationalizing the present pricing policy are offered.

Residential Water Rates: Present Practice
The present practice in water rate setting for residential consumers may be described as average pricing in theory, but as an
iterative procedure designed to recover total costs in practice.

The theoretical approach is set out in an American Water
Works Association Committee Report. [15] This report is widely
quoted in the literature; it is closely followed in a recent article
by a Canadian water supply engineer and this will be discussed

---

[14]If there are benefits from the use of small quantities of pure water by individuals that
are passed on to society in terms of health and efficiency then the subsidy is justified in
the same way that free elementary and secondary education are justified. It could be
said to be irrelevant if and only if the tax system worked to redistribute the income to the
extent that everyone would feel the cost of residential water use (or education) equally.

[15]AWWA Committee Report, "Determination of Water Rates Schedules, " Journal AWWA
46 no. 3 (1954):187-219. It was republished as Water Rates Manual (New York: AWWA,
1957).

as a typical statement by a professional in the municipal water
supply industry.[16]

> The primary function of a rate structure is to assure that the receipts
> of the water department from the sale of water and from any other sources
> shall equal or exceed the full cost of production, distribution and sale of
> water...the second general rule is that the rates be equitable i.e., the
> cost of production of water should be calculated and allocated to the supply
> of the various types of users.[17]

This indicates average pricing and in addition it makes it
necessary to allocate joint costs to the several types of users.
A recent statement in a professional journal suggests that the
rate structure should reflect the unit cost of delivering water to
the various sizes of meters; this unit cost is computed by dividing
total costs by the average amount of water sold.[18]

The shortcomings of average cost pricing will become clearer
when marginal cost pricing is discussed below; average cost
pricing satisfies the condition that enough revenues are collected
to cover expenditures but the distribution of costs among various
services and users is the problem that causes most difficulty in
establishing a rate structure. In particular the capacity costs
(i.e., those which arise out of providing plant and keeping it
ready to serve) require that some revenue be collected from the
consumer irrespective of the amount of water used. This results
in the typical 'promotional' rate structure: the first units of water
use are charged at a higher rate than the last units of water used.
However, in an increasing cost industry the last units of demand
for water often require storage facilities and cost more to produce,
but the need to distinguish between capacity and commodity costs
make it theoretically desirable to charge consumers for the right
to demand water at any time. In fact both types of costs are
required to supply water to the consumer and may be attributed
without distinction to the amount of water supplied/demanded.

[16]T.W. Argo, "Factors to Consider in Developing Water Rates Structures," in Canadian
Municipal Utilities: Waterworks Manual and Directory 1962 (Toronto: Monetary Times
Publications Ltd., 1962), p. 51 ff.

[17]Ibid., p. 51.

[18]R. P. Brock, "Concerning Services and Rates," Journal AWWA 61, no. 8 (1969):413.

The procedures suggested in the professional literature are not applied in practice. The AWWA Committee Report concludes that "the procedures for designing water rates have generally been based on trial and error assumptions applied to the sales records of the particular community."[19]

Argo notes that the rates in the 14 Ontario municipalities he examined are not comparable.[20] In 1963 the Canadian Municipal Utilities conducted an extensive survey of water rate schedules in Canada. The results were not published but a general report on the findings was drawn up by McDougall who hoped to "find a pattern or at least a positive trend in the methods used to set water rates. In fact no such pattern or trend exists. Apparently any standard can be used."[21] Later in the report it is noted that "the permutations [of water rate schedules] are infinite," and that "an objective analysis of the rates suggest that they are rarely based on logic alone."[22]

These impressions correspond with those obtained during interviews with municipal water officials in the study area. One manager described the setting of water rates as a 'hit and miss' procedure. In another municipality when the rates are charged, an iterative procedure is used until projected revenue equals projected costs, with some attention being given to the rates paid by certain user groups such as industry.

There is an increasing awareness, however, that the water supply system's obligation to meet widely fluctuating customer demand leads to overinvestment in fixed assets. The logic of incremental cost analysis is not lost upon the industry but there is the technical problem of measuring peak demand; demand meters, once they are generally available, may be the most significant factor contributing to an economically optimum price structure. For example, in one reported instance the separate metering of demand from air-conditioners and its pricing at the

[19] "Determination of Water Rates Schedules," p. 195.

[20] "Factors to Consider in Developing Water Rates Structures," p. 57.

[21] H. McDougall, "How do You Assess Water Rates?", in Canadian Municipal Utilities: Waterworks Manual and Directory, 1963, p. 50.

[22] Ibid.

TABLE 40. WATER RATES FOR METERED CONSUMERS,
31 MUNICIPALITIES IN METRO TORONTO AND REGION, 1967

Type	Frequency	Meter Service Charge	Minimum Bill Average Price with Min. Bill Equal to Commodity Charge	Minimum Bill Average Price with Min. Bill Greater than Commodity Charge	Examples
I	6	Yes	–	–	Newmarket, Acton
II	3	Yes	No	Yes	Burlington, E. York
III	4	No	Yes	No	Barrie, N. York
IV	18	No	No	Yes	Oakville, Guelph

top rate in the schedule forced the consumers to change to a
type of air conditioner that required less water.[23]

The basic elements of residential water rates that are in use
in Metropolitan Toronto and region may be conveniently
grouped under four headings:

A. Metered customers:
  i.   Fixed charge; e.g., meter service charge with or without
       a minimum bill;
  ii. Minimum bill; this includes the meter service charge where
       it applies and includes the cost of an allowance of water/
       customer/billing period at no extra cost;
  iii. Commodity charges that may have one or more (declining)
       blocks.
B. Unmetered customers:
  iv. A flat rate that may be fixed or that may vary with the
       number of rooms or the number of water-using facilities
       or the assessment value of the house.

The water rate schedules of the 45 municipalities in the study
area listed in the 1967 Annual Survey of Municipal Water Rates of
Ontario were classified; the results are summarized in Table 40
and some examples are shown in a diagrammatic form in Figures
16 and 17. The range of variations on the basic scheme makes
generalization difficult but the 31 metered municipalities are
grouped into 4 types and the unmetered 14 into two types.

[23] L. D. Kempton, "Air Conditioning Brings Water Problems," Public Works 87, no. 9
(1956):132-134.

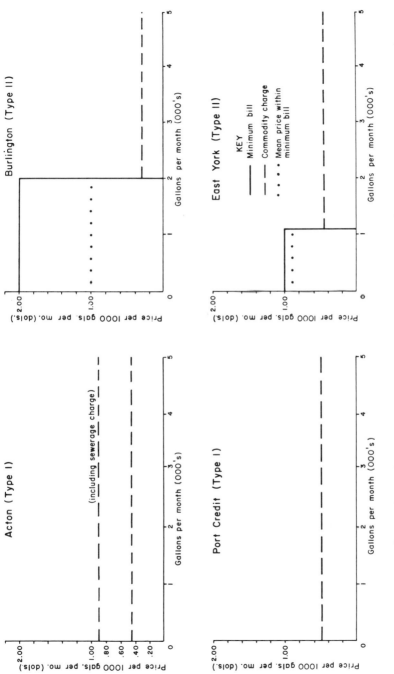

Figure 16. Types of Residential Water Rates Schedules, 1967. (Data from 1967 Annual Survey of Municipal Water Rates of Ontario, Water Works Digest [Hamilton, Ontario: Stanton Pipes, Ltd., 1967].)

The most common schedule of residential water pricing is Type IV. It consists of a minimum annual bill varying from $11.00 (York) to $41.00 (Tecumseh) that includes an annual allowance of water varying from 12,000 gallons (Waterdown) to 40,000 gallons (Chinguacousy). The average price within the minimum bill block is greater than the marginal price of the next block. This type of water rate schedule has two main defects: it provides the consumer with no incentive to use less water than the amount allowed with the minimum bill; the marginal price of water is lower than the average price within the minimum bill (i.e., the water rate per 1000 gallons declines which is the opposite of the supply cost function [Figure 15]).

The schedule for type III has an annual minimum bill ranging from $24.00 (Barrie) to $13.20 (North York) and a yearly amount of water ranging from 24,000 gallons (North York) to 60,000 (Barrie). The average price within the minimum bill is equal to the next block rate. The second criticism for type IV does not apply to schedules of type III but there is still no incentive to use less water than the amount allowed with the minimum bill; where this amount is not more than about 24,000 gallons this schedule gives the consumer some incentive to use water effectively.

Type I is better in this respect because the meter service charge (varying from $3.00 in Port Credit to $12.90 in Acton) does not include an allowance of 'free' water and the uniform marginal price is applied from the first 1000 gallons. Only 5 out of the 45 municipal water supply systems has such a schedule.

The schedule in type II has a meter service charge which is included in the minimum bill. The average price within the minimum bill is greater than the commodity price and the criticism made of types III and IV apply to this category too.

The 14 unmetered municipalities in the sample include 9 that have a fixed flat rate varying from $22.50 annually (Woodbridge) to $42.00 (Orangeville). The other five municipalities vary the flat rate according to assessment rate (Hamilton), number of rooms and the number of water-using facilities (Toronto), the number of taps (Uxbridge), single or multiple dwellings (Sutton), and the number of bathrooms (Bradford).

Generally speaking the water pricing schedules in the study area are successful in meeting one of the principles set out

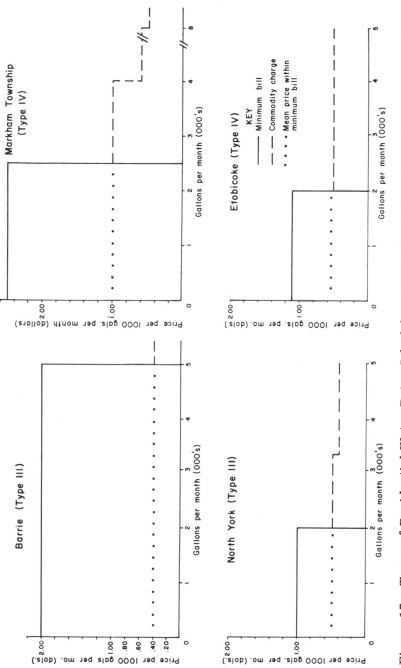

Figure 17. Types of Residential Water Rates Schedules, 1967. (Data from 1967 Annual Survey of Municipal Water Rates of Ontario, Water Works Digest [Hamilton, Ontario: Stanton Pipes, Ltd., 1967].)

above on which an ideal water rate schedule should be based, viz., collecting sufficient revenue to meet expenditures, but they fail to meet any of the other three criteria. They are arbitrary and not easy to understand; they have 'promotional' or declining block rates and the small user is discriminated against. In addition, the theory on which they are based is faulty and difficult to apply. In the next section the 'economic' solution is discussed and a water schedule that meets the theoretical and practical criteria is suggested.

RATIONALIZING THE PRICING POLICY: MARGINAL COST PRICING

In an unplanned economy, prices provide guidelines for the efficient use of scarce resources among alternative uses. In the long run consumers tend to provide enough revenue to pay for the resources used in goods and services from which they derive satisfaction. Resources are efficiently allocated among competing uses when (1) on the consumption side consumers can demand more units of the good up to the point where price is equal to the satisfaction foregone in not consuming other goods; (2) on the production side the volume of the commodity offered for sale is such that the marginal cost of production is equal to price (marginal revenue).

Resources allocated to municipal water supply would be efficiently allocated if the price of water is set equal to the marginal cost of production.[24] This would equate marginal value in use with the marginal value in exchange (price) on the consumer side and marginal cost with marginal revenue (price) on the producer side.

It has already been stated that, at least in theory, average cost pricing is followed in setting water rates; the simplest way to recover the total expenditures is to set price equal to average cost ($p = \bar{c}$) so that $pQ = \bar{c}Q$ or Total Revenue = Total Cost.

---

[24] This presentation of marginal cost pricing follows Hirshleifer, DeHaven, and Milliman, Water Supply: Economics, Technology and Policy, particularly chap. V.

Accounting or average cost price setting and marginal cost price setting provide management with the same solution only in the case where the demand curve D cuts the AC curve at its lowest point Y (i.e., where MC = AC) (Figure 18). In this case total revenue equals total cost and no 'pure' profit or loss is made.

Figure 19 illustrates the case where the demand curve (DD) cuts the AC curve to the right of Y (i.e., where AC is rising). In this case accounting (or average) price setting prescribes p_2 (< p_1) and sells OA of water. Marginal price setting results in price p_1 and quantity OB of water (<OA). OA-OB denotes the overproduction which results from setting a price that does not cover marginal cost. In this case, if marginal cost pricing is practised, total revenue exceeds total cost by (MC-AC) OB or p_1SVX.

This pure profit (or loss) is a problem of distribution and does not invalidate marginal cost pricing. One can argue that the municipal water supply system should not lose or make money. If this is the policy there may be a small allowance of free water

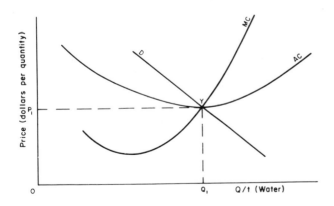

Figure 18. Solution in Range of Minimum Average Cost

for all customers so that there would be no excess revenue.[25]

Figure 20 shows the case where the demand curve cuts the MC curve in the range where average costs are falling (i. e., where MC<AC). In this case revenues are lower than expenditures and the loss could be made up by charging each customer a (small) lump sum. This 'fixed charge' would be justified in such a situation because it would enable the management to capture as revenue part of the consumer's surplus. Hirshleifer, DeHaven, and Milliman discuss other possible solutions to the recovery of the pure loss each of which "faces more or less serious objections in practice."[26]

It may be noted that the problem of allocating joint costs does not arise under marginal cost pricing; it is easy to compute the cost of reading a small number of extra meters, for example, and this would be part of the marginal cost applied to consumers for that service. If the marginal cost of serving a group of consumers differs then the price for that group should be different too.

Of the principles of setting price listed above, principle 3 is a variant of marginal cost pricing. The water rate schedule should reflect the marginal cost curve; for example, if marginal costs are rising the price schedule should have a single block, or increasing price blocks. Marginal cost pricing, on the other hand, implies either a single price or a schedule of prices such that all residential water users pay the same price for the last gallons they consume. This may create administrative difficulties.

One problem remains: should one apply the long-run marginal cost or the short-run marginal cost? Hirshleifer, DeHaven, and Milliman advocate the adoption of the short-run marginal cost.[27] Reference to Figures 14 and 15 shows that this means that prices will have to be changed up and down as the water supply plant

---

[25] The analysis outlined above assumes a regulated monopoly (i. e., it furnishes the greatest amount of product consistent with costs and demand) and not of a profit-maximizing monopoly.

[26] Water Supply: Economics, Technology and Policy, pp. 90-93.

[27] Ibid., pp. 98-99.

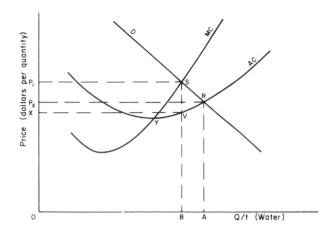

Figure 19.　Solution in Range of Rising Average Cost

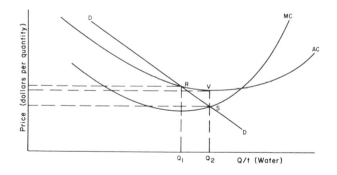

Figure 20.　Solution in Range of Decreasing Average Cost

147

grows. This may create difficulties in practice. It also means that for a series of years the water supply system will not recover expenditures unless part of the consumer's surplus is captured as revenue; then for another series of years the water supply system will yield a pure profit.

Changing marginal prices every few years would also create some confusion in the minds of the consumers; it may also be inconvenient for a family to buy a dishwasher or install an extra bathroom when the marginal price of water is low (i.e., immediately following an addition to the water supply system) and find out a few years later that it is now too expensive to use because the marginal price of water has increased considerably.

Bird and Jackson argue that the (long-run) marginal cost of the next addition should be charged because:

By the time that consumers have the opportunity to show by their reluctance to purchase from new capacity ... it is too late for the information to be taken into account [so] ... the only effective way of finding out whether a consumer is willing to pay a certain price is to charge that price now.[28]

This suggestion imposes a greater price on the consumer than is necessary. After the next addition to the water supply plant is built, the short-run marginal costs decline (Figures 14 and 15) and therefore at least for a time even the former price (equal to the long-run marginal cost preceding the addition to the water supply system) covers the cost of water. Of course the price of water may be used as a rationing device for a short period of time between the point when no more water could be produced from the existing plant to the point when the new plant comes into operation.

In practice the adoption of the long-run marginal cost as the marginal price for residential water has the advantage of being simpler to administer. Administrative procedures would be complicated if the price of water were to be changed frequently and in different directions following the short-run marginal cost curve. In addition the contract between the water supply enter-

[28]P. A. Bird and C. I. Jackson, "Economic Charges for Water," in Essays in the Theory and Practice of Pricing (London: Institute of Economic Affairs, 1967), p. 19.

prise and the residential consumers as a whole is in the nature
of a long-term contract. Once service is connected it is
reasonable to suppose that water will be on tap for the foreseeable
future. [29]

A Suggested Water Rate Schedule

The marginal price of residential water should be set equal to
the long-run marginal cost of water. One has to distinguish
between municipalities where water supply is a declining cost
industry (Figure 20) and those where water supply is an increasing cost industry (Figures 15 and 19). The former could
adopt a schedule consisting of a fixed charge and a declining
block rate; the fixed charge is required to collect sufficient
revenues to meet expenditures.

For reasons given above the cost of residential water supply
may be expected to increase over time and therefore in general
the residential water supply firm is in an increasing cost situation.
If one levies the long-run marginal cost as a single price the
revenues would exceed the expenditures all the time. The solution
is to give each consumer a free allowance of water.

The amount of water allowed free may vary in order to reflect
the short-run marginal cost. Immediately following an addition
to the system the short-run marginal cost declines and the free
allowance may be increased to say 70 gallons/day/dwelling unit.
As demand approaches capacity the short-run marginal cost
increases and the allowance of free water may be decreased to
say 30 gallons/day/dwelling unit. This allowance may be
temporarily reduced to zero when the long-run marginal cost
is equal to the short-run marginal cost (point M in Figure 15).
This would ensure that the next increment to the plant is justified
in terms of efficient resource allocation. At the same time this
flexible approach would enable the consumer to enjoy the use of
more and cheaper water when the average costs are declining
and it would keep the marginal price steady and approximately
equal to the long-run marginal cost.

---

[29] For a different viewpoint see Hirshleifer, DeHaven, and Milliman, Water Supply:
Economics, Technology and Policy, p. 97.

TARGET VARIABLES

Residential water demand management aims at the conservation of water use as a means to attaining the goal of more efficient use of the resources that the community allocates for the provision of residential water of the required standards. By influencing water demand, management can affect the level and timing of additional plants required to meet the demand. It is therefore useful to examine the allocation of investment costs in the municipal water supply industry that are designed to meet various types of demand, and to estimate the proportion of the total investment costs that are affected by a policy designed to reduce a particular type of demand.

Allocation of Investment Costs

Residential water use may be conveniently divided into several components or measures. From a policy point of view one should divide water use in a way that reflects the composition of the investment and operating costs; in the present discussion only investment costs are considered because they dominate the cost structure and are the most crucial to the decision-maker.

Five components or measures of municipal water use may be distinguished:

1. average use over the whole year in gallons/day/dwelling unit;
2. average use over the winter period (g/d/du);
3. average use over the summer period (g/d/du);
4. maximum day use (g/d/du);
5. peak hour use (g/d/du).

The winter use may be considered the basic quantity of water used by the household for the common domestic purposes; the summer use includes the winter use and the other uses of water that occur during the summer (e.g., lawn watering) while the maximum day use includes additions to the summer use.

The winter use is not significant for planning/policy purposes because this use is not likely to require increments to the water supply facilities. Data on peak hour use is difficult to obtain for individual users unless special meters are used. The most significant single measure for the design of municipal water supply systems is the maximum day use plus fire flow. [30]

Howe and Linaweaver link the capital expenditure to develop basic resources with the average daily use over a year; the expenditure to provide mains transmission and treatment facilities, distribution pumping stations, and major feeder mains is linked with use on the maximum day; the expenditure to provide local distribution mains, connections, and local storage is linked with peak-hour demand or maximum day demand plus fire flow. In the eastern United States the total system investment is allocated as follows:

30 per cent to develop basic resources;
20 per cent to build mains transmission and treatment facilities;
50 per cent to build local distribution mains and local storage. [31]

If a policy variable (e.g., price) affects maximum day demands, it would affect up to 70 per cent of the investment required to provide a residential water supply system; similarly if a policy affects average demand it would affect the remaining 30 per cent.

According to another estimate "the major investment with regard to plant cost is transmission and distribution which can be 45% to 75% of the total while supply will cost from 8% to 20% and pumping and treatment 10% to 25%. "[32] This agrees with the estimates presented by Howe and Linaweaver above if supply, pumping, and treatment are roughly equivalent to development of basic resources.

Fair and Geyer point out that:

The per capita investment in physical plant depends upon many factors: the relative proximity and abundance of a suitable water source; the need for water treatment, the availability and cost of labour and materials, the size and requirements of the system, the habits of the people and the characteristics of the area served. [33]

[30] F.P. Linaweaver, Jr., J.C. Geyer, and J.B. Wolff, A Study of Residential Water Use (Washington DC: GPO, 1967), p. iii and MacLaren (Associates), Report to the Public Utilities Commission of London on Waterworks Development to 1985, p. 7.

[31] Charles W. Howe and F.P. Linaweaver, Jr., "The Impact of Price on Residential Water Demand and Its Relation to System Design and Price Structure," Water Resources Research 3, no. 1 (1967):15 ff.

[32] S.H. Bogue, "Financial Management of a Water Utility," Journal AWWA 60, no. 3 (1968):268.

[33] G.M. Fair and J.C. Geyer, Water Supply and Waste-Water Disposal (New York: Wiley, 1965), p. 53.

They allocate 1/3 of the investment costs to collection and transportation work, 1/2 to distribution works and 1/10 to purification works.

In Ontario the share of distribution (mains, local, and local storage) investment seems to be lower than the estimates quoted so far but the evidence is somewhat tenuous. In Metropolitan Toronto the local distribution accounts for less than 22 per cent of the estimated value of the water utility plant.[34] This is the ratio owned by the boroughs who are responsible for local distribution and part of the mains distribution.

As the quote from Fair and Geyer suggests, the allocation depends upon local conditions. For London, Ontario, the following estimates are available:[35]

Source of Supply	Distribution Costs Per Cent	Development Costs Per Cent
Wells	55	45
Wells and Fanshawe Lake	38	62
Wells and Lake Huron	33	67
Lake Huron	29	71

It seems that a 50 per cent allocation to distribution is about right for a well supply but that a 35 to 40 per cent allocation should apply to a surface supply in southern Ontario. Not all investment costs are affected if the demand for water is reduced; before that topic is discussed it is useful to state what measures of demand are available and their likely impact on the demand for residential water on the maximum day.

[34] Ontario Department of Municipal Affairs, 1967 Annual Report of Municipal Statistics (Toronto: Queen's Printer, 1968), p. 186.

[35] C. B. Haver and J. R. Winter, Future Water Supply of London: An Economic Appraisal (London, Ontario: Public Utilities Commission, January 1963), Appendix A, p. 1, and Engineering Board of Review, City of London Report on Water Supply, pp. 19-20.

Selection of the Target Variables

Data and fitted equations are available from the present study regarding the impact of price and the effect of other variables on the level of water use during the year. The foregoing discussion indicates that the reduction of this variable may be applied directly to about 30 per cent of the investment cost.

The target variable of highest significance to investment decisions, water use on the maximum day (WUm), cannot be directly affected by price changes under present conditions of metering and billing even if a measure of WUm for each consumer were available. It is possible for a consumer to have a low average use during the summer but a high water use on a particular day of summer.

Although an obvious and direct link between WUm and pricing cannot be established, it is generally true that the level of water use during the summer includes most of the winter uses plus some others (e.g., sprinkling) and the water use on a hot, dry summer day includes the water uses during an average summer day plus some others (e.g., extra shower, more intensive lawn watering).

In a previous chapter it was shown that the demand curve for residential water during the winter is below that for summer in the observed price range. The fitted curves are shown in Figure 12 while the more general theoretical case is illustrated in Figures 11(a) and 11(b).

The argument may be extended by adding another demand curve as in Figure 21. At a given price p_0 the quantity demand is greatest on the maximum day (WUm) because there are additional uses of water. The demand curves are drawn such that the impact of a price change is greatest on WUm and least on WUw (Figures 21 and 22).

The empirical evidence confirms this expectation. Howe and Linaweaver report that the price elasticity is -0.225 for domestic water use and -1.16 for summer sprinkling demand. The coefficients obtained in the present study are -1.07 for summer use and -0.754 for winter use.

It follows that the impact of a price increase on summer water use (WUs) is a conservative estimate of the likely impact on maximum water use. Secondly, price is a selective instrument that affects the marginal (or extra) uses that contribute to the demand on the maximum day and that necessitate additional expenditure to increase the capacity of the water supply plant.

153

Proportion of Investment Affected

Residential water use is not the only demand that the municipal water supply system is designed to cope with. Howe[36] assumes that residential water use accounts for about 40 per cent of municipal water use and then estimates that the proportion of the investment in municipal water use allocated to residential users is about 60 per cent:

Local distribution and storage	0.70 x 0.50 or 0.35 of total municipal costs
Mains transmission and treatment	0.60 x 0.20 or 0.12 of total municipal costs
Basic source development	0.45 x 0.30 or <u>0.135</u> of total municipal costs
	0.605

It is useful to keep in mind that if residential water demand is reduced there are other parts of the system that may be un-affected but the fraction of total costs would depend on local conditions (e.g., ratio of residential demand on the maximum day to the demand of other users, the share the distribution costs account for, the relative importance of fire demand). A large suburban area is taken as an example below; in such an area residential water demand would account for most of the total demand. In the rest of the discussion the proportion of investment required to meet residential water demand only is taken into consideration; in some cases this may be 100 per cent of the total cost, in some others it may be much lower (e.g., in a small industrial town).

The development of sources, the provision of pumping and purification facilities, and the provision of local storage to meet peak demands are affected directly by a reduction in the demand for water. These items make up 18 to 45 per cent in Bogue's estimates and about 45 per cent in those of Geyer and Fair. The evidence from Ontario quoted above suggests that for pipeline

---

[36] Charles W. Howe, "Municipal Water Demands," in <u>Forecasting the Demand for Water</u>, eds. W. R. D. Sewell and B. T. Bower (Ottawa: Queen's Printer, 1968), pp. 66–68. Howe estimates that through more accurate forecasting "the total saving in system expansion investment is roughly estimated to be in the neighbourhood of $20 to $25 per capita for new populations being served" (p. 68).

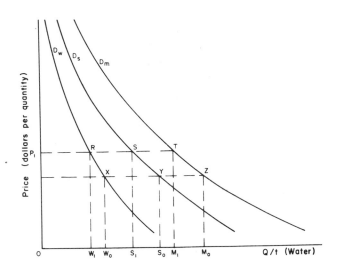

Figure 21. Demand for Residential Water on
Average Winter Day, Average Summer Day,
and Maximum Day

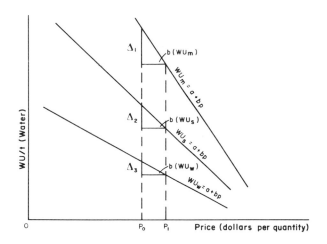

Figure 22. Demand for Residential Water on
Average Winter Day, Average Summer Day,
and Maximum Day

155

supplies the main supply accounts for about 70 per cent of the total costs, for well supplies the main supply accounts for 45 per cent and for nearby surface water supplies the main supply accounts for 50 to 60 per cent of the total costs. This median range is adopted as a 'best guess' in Table 41 below. The extent to which investment expenditures on mains transmission and local distribution are affected is less clear. The amount and spatial distribution of the water pipe system under the streets reflects the size and distribution of the population served.

There are two ways that the level of such investment may be affected by the level of demand, but only in a marginal sense. The main cost component of putting in water pipes (mains or local) is the cost of labour and this does not vary with the size of pipes; there are small differences in cost of material and labour (e. g. , depth of ditch required) between laying a 6" or a 12" pipe and these costs may be affected by a reduction in the design capacity of the system or sub-system serving a residential area.

Secondly, some mains have to be replaced after a time because they cannot cope with the increased capacity demanded in the area. Some of this increased capacity is due to an increase in the built-up area but some is due to an increased rate of water use per residence. A reduction in the rate of water use will postpone some of these replacements.

In conclusion, one may reasonably expect that about two-thirds of the investment requirements for residential water supply are affected by policy changes that reduce the rate of water use per dwelling unit. These are distributed as shown in Table 41.

TABLE 41. PROPORTION OF CAPITAL INVESTMENT IN MUNICIPAL WATER SUPPLY AFFECTED BY A REDUCTION IN THE DESIGNED CAPACITY

Facility	Design	Allocation of Investment Per Cent	Per Cent of Total Affected by Policy	Target Variable
Basic resources	WUa/WUm	30		WUa/WUs
Treatment, pumping, storage	WUm	20 – 30	50 to 60	WUs
Mains, transmission and local distribution	WUm	40 – 50	10	WUs
		100	60 to 70	

INSTRUMENT VARIABLES

This section deals mainly with price as a management tool. This
is not the only policy instrument available to the management of
municipal water; in Chapter II (Table 17) several policy
alternatives were listed and grouped according to the degree
of change required from the policy-makers. Another convenient
classification of these alternatives would be under four headings:
(a) increasing supply, (b) using water produced more efficiently
in a physical sense, e.g., waste water reclamation, preventing
leaks from mains; (c) 'shifting' water resources from some uses
to others (e.g., agriculture to municipal) or from one supply
plant in one municipality to another municipality; (d) demand
management through metering and pricing, or through regulation
and police action. Some non-pricing alternatives are treated in
outline in the next section.

Pricing as a Policy Variable
The present pricing policy leads to overinvestment if the marginal
price of water is less than its marginal cost. On the whole water
rate schedules reflect the tacit assumption that the level of water
use is not affected by price because (a) water is cheap, (b) water
is essential to many household and personal needs, and (c) it is
a small item in the family budget. In a previous chapter it was
shown that residential water users are responsive to price
differences and the preceding section provided some background
to more logical pricing procedures. These alternatives are
applied to a 'new' city of 200,000 people using some of the
empirical findings reported above in Chapter III.
    A new city of 200,000 people is taken as a 'model' in order
to simplify three difficulties: changes over time, fire demand,
economies of scale.

1. This study is concerned with spatial differences or cross-
sectional differences. These reflect long-run differences and
the findings can be applied, strictu sensu, only over space
rather than over time. In a more speculative interpretation the
findings are indicative of the direction and amount of change in
the level of residential water use over a period of time sufficiently
long to allow the consumer to make adjustments (e.g., get used

to less frequent lawn watering or to having faulty equipment attended to).

2. The investment in water supply facilities required for fire protection can represent a quarter to three-quarters of the total investment, the proportion being inverse to the size of the population (Table 42). In small communities the effect of higher prices on the storage requirement would be swamped by the need to cope with fire demand.

3. In constructing water supply facilities there are important economies of scale; hence the large size of the community would simplify the presentation of the policy alternatives. One other simplification is introduced into the model; the 200,000 people live in single-unit residences with an average size of family 4 so that there are 50,000 residences. The fixed bill is set at a low level, 50 cents/month or $1.50 per billing period; the fixed bill reflects present practices and may be removed under an improved pricing system.

TABLE 42. STORAGE REQUIRED TO MEET FIRE DEMAND

Population	Flow Rate Gallons/ Minute	Duration of Pumpage (Hours)	Required Storage (Million Gallons)
5,000	2,250	9	1.2
10,000	3,000	10	1.8
40,000	6,000	10	3.6
100,000	9,000	10	5.4
200,000	12,000	10	7.2

SOURCE: AWWA Committee Report, "Determination of Water Rate Schedules," Journal AWWA 46, no. 3 (1954):202, table 5.

Table 43 shows the expected residential water use on a summer day (WUs) for this community. The water use for metered households is obtained by using the fitted equation (28) above. Only summer use is shown because it is the surrogate for the design variable, maximum day use (WUm).

The observations of unmetered water use discussed in Chapter III refer to townhouses and therefore underestimate the summer water use in single-unit housing which has a larger lawn area. The summer water use (WUs) for a townhouse with four persons in residence is derived from equation (32):

Value of Residence $	Summer water use (gallons/day per dwelling unit)	Increase in summer water use per $5,000 of residence value
20,000	193	
25,000	225	32
30,000	256	31
35,000	285	29
40,000	313	28

The average summer use (i.e., June, July, and August) in Etobicoke for single-unit non-metered residences is 68.6 gallons/capita/day[37] or 274.4 gallons/day for a household of 4 persons, and an assessed sales value of about $19,000. One can adopt this figure for the $20,000 group of unmetered residences and increase it by the figure in the third column in the table above in order to get an estimate of water use in the other groups. These figures are the ones in Table 43.

Metering

In unmetered residences the marginal price of water to the consumer is zero. There is no incentive to use water efficiently particularly for lawn-watering. Table 7 shows that non-metered consumers in the West use 2.6 times the lawn watering requirement on their lawn while for metered consumers in the West the factor is 0.62 and for metered consumers in the East it is 0.42. There is also no incentive to discover and repair faulty water fixtures such as closet tanks when water use is not metered.

Enough evidence is reported in the literature to show that non-metered consumers use about 50 per cent more water than

---

[37] Borough of Etobicoke, Engineering Department, "Etobicoke Study." Mimeographed, n.d.

TABLE 43. ESTIMATED SUMMER AVERAGE RESIDENTIAL
WATER DEMAND IN A 'NEW' COMMUNITY OF 200,000

Value of Dwelling	Water Use per Dwelling Unit (Gallons/Day)	Number of Dwelling Units	Total Water Use (Million Gallons Daily)
Unmetered, marginal price = 0			
20,000	274.4	15,000	4.116
25,000	306.4	15,000	4.596
30,000	337.4	10,000	3.374
35,000	366.4	5,000	1.832
40,000	394.4	5,000	1.972
			15.890
Metered, marginal price = 30 cents/1000 gallons			
20,000	287.73	15,000	4.316
25,000	322.62	15,000	4.839
30,000	354.26	10,000	3.543
35,000	383.41	5,000	1.917
40,000	410.59	5,000	2.053
			16.668
Metered, marginal price = 40 cents/1000 gallons			
20,000	211.49	15,000	3.172
25,000	237.14	15,000	3.557
30,000	260.39	10,000	2.604
35,000	281.82	5,000	1.409
40,000	301.805	5,000	1.509
			12,251

metered consumers in the long run and that most of this increased use occurs during the summer. Whether consumption is reduced following metering depends on the pricing structure adopted, particularly the level of the marginal price.

Table 44 shows the effect of metering and marginal price on water use and investment for the hypothetical new community of 200,000 people; it is assumed that the per capita investment for the provision of municipal water is $300 (1965 dollar value, see Table 2) and that half of this is allocated to residential water supply.

160

Table 43 (continued)

Metered, marginal price = 50 cents/1000 gallons

20,000	166	15,000	2.490
25,000	187	15,000	2.805
30,000	205	10,000	2.050
35,000	222	5,000	1.110
40,000	237	5,000	1.185
			9.640

Metered, marginal price = 60 cents/1000 gallons

20,000	137	15,000	2.010
25,000	153.7	15,000	2.3025
30,000	168.7	10,000	1.685
35,000	182.5	5,000	0.9125
40,000	195.0	5,000	0.9775
			7.9325

Metered, marginal price = 70 cents/1000 gallons

20,000	116.2	15,000	1.7430
25,000	130.0	15,000	1.9500
30,000	143.0	10,000	1.430
35,000	155.0	5,000	0.775
40,000	165.5	5,000	0.8275
			6.7255

Metered, marginal price = 80 cents/1000 gallons

20,000	100.74	15,000	1.511
25,000	112.96	15,000	1.694
30,000	124.03	10,000	1.240
35,000	134.24	5,000	0.671
40,000	143.75	5,000	0.719
			5.8350

Unlike other policy alternatives open to water management the introduction of metering is not a costless alternative;[38] there are costs for the purchase and installation of meters and the recurrent cost of meter reading and maintenance. In addition there is a loss in the consumers' surplus, i.e., the total area under the demand curve less the area above the price line.

[38]The theoretical formulation of the optimal timing for the introduction of metering is discussed by J.J. Warford, "Water Requirements: The Investment Decision in the Water Supply Industry, " The Manchester School of Economic and Social Studies 34, no. 1 (1966): 87-112.

TABLE 44. THE EFFECTS OF METERING ON RESIDENTIAL
WATER USE (Average Summer Day, WUs)

Marginal Price (Cents/1000 Gallons) (i)	Reduction in WUs from Unmetered		Reduction in Investment		
	Million Gallons/Day (ii)	Per Cent (iii)	(iii) X 65 Per Cent (iv)	Per Capita Dollars (v)	Total Million Dollars (vi)
40	3.65	22.97	14.93	22.395	4.479
50	6.25	39.33	25.56	38.35	7.675
60	7.96	50.1	32.55	48.83	9.765
70	9.165	57.68	37.49	56.233	11.247
80	10.05	63.25	41.11	61.666	12.333

NOTES: (a) Total investment required is $300 X 200,000 = $60.0 million.

(b) The percentage of investment requirement for residential purposes is assumed to be 50% of which 65% is assumed to be affected by a reduction in design capacity.

The benefits are (1) the reduction in the operating costs due to the decreased demand and (2) the saving in the investment required to meet water demands that have been obviated through metering. In Figure 23 the curve marked Df is the demand for residential water over time if the water use is not metered; Dm shows the demand for residential water over time if the water use is metered. The decision to introduce metering is examined in t_0. Metered consumption is OA, unmetered consumption is OB. The difference, AB, is the reduction in water use and there is a saving of costs associated with the production and distribution of this amount of water. The management is also able to postpone further investment in the water supply system until t_1; this benefit is equal to the present value of the additions that would have been built to provide the quantity CB of water. Alternatively, if the system is not yet built, there is a saving in investment which is equal to the present value of the investment required to increase water supply by AB; the associated operating costs are also avoided.

Two studies of the merits of introducing metering for residential water consumers have been carried out by municipalities in the study area; an evaluation of these studies is useful because

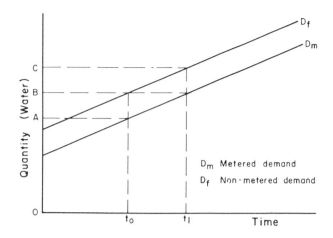

Figure 23. Benefits of Metering Residential Water Use

it indicates how present practice differs from the procedure outlined above. These studies also bring out a couple of interesting issues and they provide an indication of the costs and benefits of metering in southern Ontario.

City of Toronto study
In the city of Toronto most residential buildings are old and metering would involve some extra cost in inspection and installation. All apartments larger than four-plexes have been metered since February 1966, i.e., prior to the benefit-cost study carried out by consultants and the Department of Public Works.[39]
One should also note that the benefit of postponed investment does not accrue to the City because water is bought from the Metropolitan Corporation at a flat rate of 25 cents/1000 gallons. Storage and effluent treatment is provided by Metro and therefore the incentive to reduce peak demands by the City water management is weakened. On the other hand, the peak demand

---

[39] City of Toronto, Department of Public Works, Report on City of Toronto Water Distribution System (Toronto: April 1968), and James F. MacLaren (Consulting Engineers), Report and Technical Discussion to Accompany the Report on the Water Distribution System for the City of Toronto (Toronto: January 1968).

163

ratio in central cities are normally lower than in the residential suburbs.

Meters reduce water consumption and consequently reduce the cost of water by minimizing water and waste treatment, pumping and storage. [This] is generally valid in a city that is responsible for water and waste treatment. However in Toronto this is not so because the Metropolitan Corporation provides the city with treated water and also treats the associated waste. [40]

The Report on City of Toronto Water Distribution System recommends that the present policy of metering new apartments larger than four-plexes and all industrial and commercial users be continued; that existing nonmetered apartments larger than four-plexes be metered; and that the present flat rate structure for residential consumers be reviewed at least every five years. [41] The decision was taken on the basis of a benefit-cost study; the question of equity among consumers was also examined. The "cost of water on the average is the same to a flat-rate customer as to a metered customer." [42]

The benefit-cost study as described in the Report and Technical Discussion is set out below; some changes in the method and assumptions are introduced later on to examine whether the outcome would have been different.

The annual costs of introducing 100 per cent metering are (in 1966 terms): [43]

Debenture financing for meter purchase and installation	$560,904
Ten additional meter readers	75,000
Meter maintenance	70,000
Clerical expenses	20,000
	$725,904

[40] MacLaren (Consulting Engineers), Report and Technical Discussion, p. 100.

[41] City of Toronto, Department of Public Works, Report on City of Toronto Water Distribution System, p. 28.

[42] Ibid., p. 4. Since non-metered do not require meters and their associated costs, their water should be cheaper.

[43] MacLaren (Consulting Engineers), Report and Technical Discussion, pp. 116-117.

The annual benefits, defined as the net change in revenue are:[44]

Decrease in expenditure for water purchase	$317,932
Decrease in revenue (present flat rate revenue less revenue from metered residences)	303,618
	14,314

This makes the annual costs exceed the benefits by over $700,000. There are two aspects of the analysis that invite comment. The first is that in the present instance the City has to decide whether the annual savings in water purchases are more/less than the costs of metering. The figures above show that metering is not recommended because the costs are about double the benefits.

The second aspect is the assumed reduction in water use, i.e., the magnitude of the benefits. The <u>Report and Technical Discussion</u> refers to the Johns Hopkins Residential Water Use Study and comes to the conclusion that "in the City of Toronto flat rate consumers may be assumed to use approximately 15 per cent more water than comparable metered consumers."[45] In Table 45 are the reported findings of the Johns Hopkins Study. These figures suggest that the increase in water use from metered to unmetered consumers is 51 per cent on an average day use and 140 per cent on the maximum day use. The average day use rate is the relevant one for the city. The data collected by the Borough of Etobicoke Engineering Department shows an increase of 45 per cent in water use of unmetered customers living in streets of comparable assessment values. The benefits are worked out in Table 46 assuming different rates in the reduction of water use; the costs are those given above, $725,904. If the ratio of metered to unmetered water use is assumed to be 100:145 the savings to the city community in operating costs would exceed the annual charges of metering.

Another decision open to the city water management is to meter newly constructed residences. The costs of inspection and fitting the meter in an old connection would be obviated in newly con-

[44] Ibid., pp. 117-120.
[45] Ibid., p. 38.

165

TABLE 45. COMPARISON OF METERED VERSUS FLAT-RATE
USE IN THE WEST OF THE UNITED STATES

	10 Metered Areas	8 Flat-Rate Areas
	(Gallons/Day/Dwelling Unit)	
Average annual	458	691
Maximum day	979	2354
Peak hour	2481	5171

SOURCE: F. P. Linaweaver, Jr., J. C. Geyer, and J. B. Wolff,
A Study of Residential Water Use, p. 50, table 6.

structed residences; the costs of meter reading and maintenance
due to a few additional metered customers would be minimal and
the data collected would be useful when the flat-rate structure is
reviewed. The benefit and cost comparison in this case would be
reduced to comparing the reduced cost of purchasing water and
the financing cost of meter purchase and installation. Let us
assume that there will be 100 dwelling units metered annually
under this policy. The cost of installation and purchase is
$3,900.00 at 7 per cent over 20 years, i.e., $368 per annum.
The reduction in the cost of water purchases at $0.25/1000
gallons is $226 assuming a reduction of 13 per cent. The rate
of reduction that would make benefits equal to costs is 27.5/
127.5, which is well below the 45/145 reduction rate required
to make metering of all old residences worthwhile.

Etobicoke study
In 1968 and 1969 the Engineering Department of the Borough of
Etobicoke examined the economic feasibility of installing meters
in New Toronto, a municipality annexed in 1967. There are
2018 consumers in this area and they make up 3.8 per cent of
total consumers in the borough.

The Etobicoke study differs from the Toronto study in two
respects:

1. The cost of metering other than the debenture financing are

TABLE 46. REDUCTION IN EXPENDITURE ON WATER PURCHASE
FOLLOWING METERING, TORONTO, 1966 (Assumed Figures)

Assumed Ratio Metered:Unmetered Water Use	Reduction in Residential Water Use (Million Gallons)	Reduction in Bill for Water Purchase $
100:115	1271.728	317,932
100:120	1624.986	406,246
100:125	1949.983	487,496
100:130	2249.981	562,495
100:135	2527.756	631,939
100:140	2785.691	696,423
100:145	3025.836	756,459
100:150	3249.962	812,490

not estimated since the meter reading staff can absorb the extra
work without additional expenses.

2. The consumer is assumed to maintain the level of water use
following metering; the level of water use in non-metered residences is about 45 per cent higher than in comparable metered
residences.

The benefit-cost analysis as reported is set out below:

	$
Consumption per quarter in unmetered residences: 15,950 gallons	
Revenue on a flat-rate basis	5.70
Revenue on a metered rate basis	8.93
Increase in revenue following metering (assuming that consumption does not change) $3.23 x 4 x 2018	26,072.56
Capital and installation costs	125,116.00
Annual cost of financing debenture over 7 years at 8.5%	24,625.00
Annual cost of meter maintenance	1,250.00
Total annual costs	25,875.00
Total annual benefits (increased revenue)	26,072.00

The incentive for the borough to meter the water use in the residences not already metered is very clear. There is the added argument of 'equity,' i.e., these residences would now be treated like other consumers.

If the level of water use were to decline by the full difference between metered (11,000 gallons per quarter per residence) and non-metered (16,000 gallons) the increase in revenue would be $6,054.00. The decision to introduce metering is justified if the debenture is financed over 20 years[46] as the following figures indicate:

Saving on bill for water purchase	$	9,989.00
Increased revenue		6,054.00
Total 'benefits'		16,043.00
Annual cost of debenture financing over 20 years at 8.5%		13,220.00
Annual cost in meter maintenance		1,250.00
Total cost		14,470.00

If the decrease in the cost of water produced is compared with the increased costs due to metering, then the introduction of metering is not an efficient solution. One possibility is to increase the flat-rate charge to cover the cost of purchasing water plus overheads and give the consumers the option to have meters installed.

The average cost of water in Metropolitan Toronto in 1967 was 34 cents/1000 gallons and the cost of water used during peak days for lawn watering (which is the particular use that metering affects most next to 'waste') must be higher. If Metro levied a summer charge for extra heavy demands, the reduction in the water bill for the municipalities would make the extra expense of metering worthwhile. In addition there would be investment costs that could be postponed. The alternative to the introduction of metering is very relevant to the study area:

[46] Meters are considered to have a useful life of 20 years; for example see Warford, "Water Requirements: The Investment Decision in the Water Supply Industry," p. 103.

there are about 15 municipalities that do not have 100 per cent metering, including the City of Toronto and Hamilton.

In terms of investment costs the case for metering is clear: Table 44 indicates that the reduction in investment requirements range between $22 and $62 per head depending on the marginal price level; the Etobicoke study indicates that the introduction of metering required $62 per residence.

Rationalizing the Water Rate Schedule

The water rate schedule in general use may be improved upon in three ways:
1. by abolishing the minimum bill including any demand charges such as meter service charges;
2. by abolishing the allowance of water with the minimum bill;
3. by increasing the marginal price (commodity charge) so as to include the cost of treatment of waste water from residences.

In some municipalities all the changes could be implemented at the same time (i.e., where the marginal price is greater than the average cost) but in others the minimum bill cannot be abolished altogether (i.e., where the average costs are declining so that efficient resource allocation would result in a marginal commodity charge lower than the average cost). For this reason it is proposed to discuss the alternatives in turn.

The fixed cost of water supply are relatively high mainly because of the large investment required; in addition, some of the operating costs are also fixed irrespective of the amount of water sold (e.g., meter reading and billing). The minimum bill is designed to cover some of these costs and so raise the necessary revenue.

The fixed service charge or the minimum bill is justified only if the short-run marginal costs are lower than the average costs. In this case charging the average cost would not cover the total expenditure and therefore it is necessary to capture part of the consumers' surplus in order to raise the necessary revenue.

When marginal costs exceed average costs, minimum bills are not justified and are wasteful of resources. It is not considered necessary to pay a fixed annual charge to be able to buy milk or bread during the year; fixed capital and operating costs, like the variable costs, are reflected in the price charged for the commodity.

The minimum bill results in a declining block rate for all residential consumers. This is a promotional rate: the more one consumes, the lower the price. It encourages the use of water by some consumers, viz., those who can afford to consume more and are most likely to contribute to peak demands.

Twenty five of the 31 municipalities where residential water use is metered (Table 40) have a minimum bill. In 21 instances (Types II and IV) the minimum bill is equivalent to having a declining block rate. Table 47 classifies these municipalities into those that have a commodity or marginal price greater than the average cost (7) and those that charge a marginal price less than the average cost (12); information is not published for the two others.

The 12 municipalities whose commodity charge is less than the average cost may be in the part of the average cost function where it is declining (Figure 20); if the minimum bill is abolished, there will not be sufficient revenue to cover the expenditure. These municipalities would simplify the water rate schedule and make it more rational by abolishing the allowance for water with the minimum bill and retain a minimum charge; their revenue could only increase and the minimum bill could be progressively reduced.

Consumption of water would be reduced if the allowance with the minimum bill is discontinued (Figure 10). For example, Burlington has a minimum bill of $24.00 per annum and allows 24,000 gallons, Markham Township allows 30,000 gallons with a minimum bill of $30.00 per annum and Oakville allows 36,000 gallons with a minimum bill of $29.40 per annum.

The declining block rate effect of the minimum bill cannot be justified when enough revenue can be collected through the commodity charges. This seems to be the case in those munici-palities where marginal price is greater than average cost. [47] The abolition of the minimum bill may require an increase in the marginal price of water; in other municipalities the excess revenue may make it possible to give away a small quantity of water (say 1000 gallons per month) to each consumer.

---

[47] If marginal cost is below average cost the price should be reduced since the consumers are paying more than the marginal cost.

TABLE 47. DIFFERENCES BETWEEN MARGINAL PRICE AND AVERAGE
COST OF WATER IN MUNICIPALITIES WHERE MINIMUM BILLS APPLY

Number of Municipalities	Marginal Price less Average Cost 1967 (Cents/1000 Gallons)			Examples
1			-100 or less	Markham Township
4	-50	to	- 99.9	Vaughan Township, Ancaster
7	0	to	- 49.9	Burlington, Brampton
4	0	to	+ 9.9	Etobicoke, Ajax
3	+10	to	+ 19.9	E. York, York, Scarborough

The impact of abolishing the allowance of water with the minimum bill on the level of residential water use will be small if the amount allowed is 1000 gallons or 2000 gallons per month/residence because most consumers exceed such consumption levels. There are ten municipalities in the study area where the amount of water allowed with the minimum bill exceeds 2000 gallons per month/ residence because most consumers exceed such consumption levels. There are ten municipalities in the study area where the amount of water allowed with the minimum bill exceeds 2000 gallons per month/residence and in these municipalities the reduction in water use following the abolition of the minimum bill may be considerable. There is no coefficient that one can use from this study; equation (14) indicates that the elasticity of water use with respect to this variable is 0.5, i.e., if the amount of water with the minimum bill is reduced by 10 per cent the consumption of water would be reduced by 5 per cent. However, this coefficient is derived from a sample where the amount of water allowed with the minimum bill is low and the standard error of the coefficient is relatively large. Therefore the matter requires further investigation.

The marginal price for water should at least be equal to average cost; in most municipalities in the study area the abolition of the minimum bill would require an increase in the commodity charge for water and the impact of this instrument variable is relevant to the allocation of resources to residential water supply over most of the area.

When the water charges for residential consumers are increased the management may increase the commodity charge, or the minimum bill, or both. The water supply management of the hypothetical city of 200,000 may adopt a wide range of prices under present practice. If the average cost of water is 60 cents/1000 gallons, then a two-month bill for 8000 gallons could be collected in a number of ways, for example:

1. A minimum bill of $4.00 for 4,000 and a commodity charge of 20 cents/1000 gallons;
2. A minimum bill of $3.00 for 3,000 gallons and a commodity charge of 36 cents/1000 gallons;
3. A minimum bill of $2.00 for 2,000 gallons and a commodity charge of 46.666 cents/1000 gallons.

Table 48 indicates the reductions in the level of water use during the summer period (WUs) and the reductions in the required investment that are possible. For example, in the study area the most common marginal prices in 1967 were between 40 and 50 cents per 1000 gallons. If one opts for 70 cents rather than 40 cents the reduction in water use during the summer period is 110 gallons/dwelling unit/day or 5.515 MGD and the reduction in investment is 30 per cent for $150 per capita or $8.785 million.

Inclusion of the Cost of Waste Water Treatment in the Marginal Price

There are a number of municipalities that levy a sewerage charge as part of the water bill. In Pickering Township the sewerage charge used to be 50 per cent of the commodity charge and the fixed bill: in Newmarket and Acton the sewerage charge is included in the commodity charge of 90 cents and 80 cents/1000 gallons respectively.

This policy is not very common among municipalities and in one instance (Pickering Township) the sewerage charge has been transferred to the general tax bill. However, the practice of a combined water-supply and treatment charge has a lot to commend it since the water used in the home, except for the water used to irrigate the lawn, has to be treated unless it is allowed to lower the quality of the receiving water. Complete sewage treatment works are relatively twice as expensive as water purification works whereas collection systems for domestic

TABLE 48. IMPACT OF MARGINAL PRICE ON RESIDENTIAL
WATER USE (Average Summer Day)

Marginal Price (Cents/ 1000 Gallons) (i)	Reduction in Water Use from Base Price Level		Reduction in Required Investment from Base Price Level		
	M.G.D. (ii)	Per Cent (iii)	(iii) X 0.65 Per Cent (iv)	Per Capita Dollars (v)	Total Million Dollars (vi)
Base price 30 cents/1000 gallons					
40	4.42	26.53	17.24	25.86	5.173
50	7.02	42.13	27.39	41.08	8.218
60	8.727	52.38	34.00	51.00	10.215
70	9.9345	59.63	38.76	58.14	11.628
80	10.82	64.94	42.21	63.32	12.664
Base price 40 cents/1000 gallons					
50	2.60	21.24	13.81	20.71	4.142
60	4.3075	35.19	22.87	34.31	6.862
70	5.5145	45.05	29.29	43.93	8.785
80	6.40	52.29	33.99	50.98	10.196
Base price 50 cents/1000 gallons					
60	1.707	17.71	11.51	17.27	3.3453
70	2.9145	30.23	19.65	29.48	5.896
80	3.800	39.42	25.62	38.43	7.687
Base price 60 cents/1000 gallons					
70	1.207	15.21	9.89	14.83	2.967
80	2.09	26.38	17.15	25.72	5.144
Base price 70 cents/1000 gallons					
80	0.8855	13.17	8.86	12.84	2.567

sewage are about half as expensive as distribution systems for water.[48] Besides lowering the investment requirements for water purification plant, there would be a further saving by lowering the investment requirements for sewage treatment plant.

The adoption of this policy alternative would approximately double the present commodity charge for residential water. Table 48 indicates that if the price is increased from 30 to 60

[48]Fair and Geyer, Water Supply and Waste-Water Disposal, p. 83.

cents/1000 gallons the investment requirements are reduced by 34 per cent of $30 million or by $10.2 million. About the same saving would result from doubling the price from 40 to 80 cents/1000 gallons.

Seasonal Charges

Water supply systems capacity is usually designed to meet maximum day demands with the result that some capacity is idle some of the time. The extent of capacity that is idle is indicated in Table 49 which shows the demand and its duration. The maximum demand that the capacity of the system can meet is 145 per cent of the average day demand. Slightly less than 1/2 of the capacity is required on all days of the year; about 2/3 of the capacity is required only on half the number of days. About 14.4 per cent of the capacity is required not more than once every 20 days or on about 18 days in a year; in other words one-seventh of the system is idle on 348 days a year.

There are two ways of recovering the marginal cost imposed upon the water supply system to meet peak demands. One is to fix peak-responsibility charges which will be in proportion to the load that the consumer-group places on the peak demand. For example, if a residential area necessitates a peak capacity of 200 per cent of the average capacity to meet demand, then the extra cost should be passed on to the residential consumers as a fixed charge. This method would not allow individual consumers to adjust their on-peak use according to the cost and therefore does not allow the consumers to equate marginal utility with price.

It is better to have peak-load pricing. On-peak and off-peak service may be considered as two different goods and each may be priced separately. This method would allow on-peak users to adjust their level of use in response to the marginal price (which would equal marginal cost) and so result in the efficient allocation of resources.

Ideally there should be separate meters for on-peak and off-peak water use. Such separate metering would impose a high investment cost which could be avoided by the use of summer charges. The reasons for expecting an increase in the price of residential water during the summer to reduce maximum day use have been presented above. It is enough to add that the number of 'crises', or events of short-term water scarcity during spells of hot dry weather, will be reduced and the additions

TABLE 49. DURATION OF DEMAND IN A
MUNICIPAL WATER SUPPLY SYSTEM

Duration (Per Cent of Total Days in a Year)	Demand (Per Cent of Average Day)
100	70
90	81
80	88
70	92
60	96
50	100
40	104
30	108
20	111
10	118
5	124
0	145

SOURCE: James F. MacLaren, Report and
Technical Discussion on Functional Design of
Lake Huron Supply for the Public Utilities
Commission of the City of London (London,
Ontario: PUC, February 1964).

to the plant to cope with peak demands will be postponed.

The impact of a seasonal charge would be the same as that of a price increase considered above (Table 48). There are serious objections to this alternative. All meters would have to be read on approximately the same date as the summer charge comes into effect. In most of the municipal water billing offices the work load is spread over the whole year by dividing the municipality into areas, or divisions or cycles; these areas are billed in turn. Seasonal charges would impose costs either for the purchase, installation, and maintenance of special meters or for an increase in the number of meter readers and billing personnel. The same effect of summer charges may be expected from an increase in the marginal price or from an increasing price block schedule, and they are relatively costless alternatives; for this reason the alternative of summer charges cannot be justified on the basis of the present analysis.

175

An Increasing Block Rate Schedule

If the marginal cost of providing residential water is increasing, at least in the long run, then the consumers who use more water on the average are imposing a cost on other consumers.[49] In Figure 24 the consumers who use 120 gallons/day are imposing an extra cost of $C_1 - C_2$ for the extra 20 gallons/day/consumer.

It is valid to consider the water produced at a greater cost to be a different good or service from the water produced at a lower cost. The high consumer of water should be charged a price for the extra 20 gallons/day that would equal the marginal cost C_1.

It is desirable to have a limited number of price blocks in the price schedule. In order to cover total costs the revenue has to equal $\Sigma C_i q_i$ where q_i is the quantity of water in the ith segment of water production and C_i is the cost of producing the water in the ith segment of the production function. In order to recover costs and to set a small number of prices that reflect marginal costs in the long-run one has to set a price such as p_2 in Figure 24 so that the area of the shaded parts marked A is equal to the area of the shaded part marked B. The second block would have a price p_1 such that the area shaded and marked R would be equal to the area shaded and marked S.

Resources are optimally allocated if the price is set equal to marginal cost; this condition would be satisfied if the price in the highest block is set (approximately) equal to marginal cost. In addition, this residential water rate schedule allows small consumers of water to satisfy their demand at a price that reflects the amount of resources the community has to set aside if all consumers demanded small amounts during peak periods; high users of water would have the option of paying for the resources which their high use of water at the peak makes necessary. This alternative would meet the objection put forward by the US Senate Select Committee on National Water Resources that "using price as a means of curtailing water used under these circumstances would also be inequitable since economically well-off citizens would probably continue to use all the water they wanted."[50]

---

[49] If average costs are declining in the long-run, then this alternative is not open to the management. However, in a city with a demand increasing over time and the peak demand increasing due to higher incomes, an increasing cost situation may be assumed.

[50] US Congress, Senate Select Committee on National Water Resources, Water Requirements for Municipal Use, 86th Cong., 2nd Sess., Committee Print No. 7, (Washington DC: GPO, January 1960), p. 16.

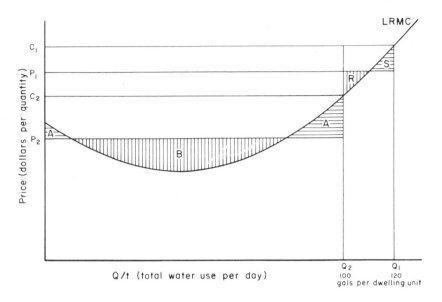

Figure 24. Long-Run Marginal Cost Pricing

If an increasing price schedule is adopted the management would be in a position to make available a small amount of water at a low cost to all consumers; those who choose to buy more will be expected to pay a price that reflects its cost of production. The present practice enables the economically well-off to buy water more cheaply than those who cannot pay for (or do not need) more water than the amount allowed with the minimum bill.

The pricing schedule suggested above (following the review of present water rate schedules) could be used in this context: the first price would be very low (or even zero) and the second price reflecting the long-run marginal cost. Where average costs are rising steeply or the summer peak is very marked, a three price schedule would be better in order to distinguish among consumers who contribute little to the peak demand, consumers who contribute moderately to peak demand and consumers who contribute a lot to the peak demand. For example, the first block would have a small allowance of cheap water (say 1000 gallons/month/dwelling unit for 30 cents). The second block could vary with

177

the short-run cost of supply; immediately following the addition of an increment to the water supply system, short-run average costs are declining (i.e., there is over-capacity). By allowing more middle-priced water the municipal water management can make use of capacity that would otherwise remain idle. The highest-priced block would reflect the long-run marginal cost of water.

The advantages of an increasing price block schedule may be briefly summarized:

1. This schedule is simple to administer;
2. It makes possible the recovery of expenditures through water revenues;
3. It (approximately) equates marginal price with the long-run marginal costs; at the same time the medium or low price block would reflect the short-run marginal cost;
4. It would make frequent changes of prices unnecessary;
5. It would serve the same purpose as summer charges; if consumers pay a higher price for water demanded during peak periods the demand on the maximum day would decrease in general;
6. This schedule takes into account the fact that the use of high-quality water by individuals for essential purposes is beneficial to society as a whole and should therefore be supplied free or at low cost.[51]

The reduction in water use to be expected from this alternative is probably of the same order as the reduction following increases in the marginal price (Table 48). An interesting feature of such a water rate schedule would be that the consumer would still enjoy a large proportion of the consumer's surplus.

INVESTMENT VARIABLES: NON-PRICE ALTERNATIVES

In considering ways of managing the demand for residential water first priority should be given to metering and a rational water rate schedule because these alternatives do not impose any

---

[51]The relevance of such a feature to areas where the income of the masses is low should be emphasized.

artificial restrictions on water use. However, when the marginal price is relatively high further increases become progressively less effective. The following non-price alternatives are therefore more attractive when the projected marginal price in the design water supply system is already relatively high.

Reduction in the Capacity of Domestic Plumbing Fixtures

About 16 gallons of water per capita per day are required to flush away toilet wastes. The capacity of flush tanks in Toronto is about 4 gallons; if this could be reduced there would be a proportional decrease in water use from the total of 3.2 MGD for the city of 200,000 taken as a design system. There are models on the market that use only 1/4 gallons of water per flush; there may be difficulties with the sewage system if the amount of water in the effluent is reduced drastically but if water becomes expensive to produce, this alternative may be considered. When the marginal price is 50 cents per 1000 gallons, the average residential water use on a summer day is 9.64 MGD (Table 43). If the capacity of flush tanks is reduced by 50 per cent the water use is decreased by 1.6 MGD or 16 per cent. This is comparable to the reduction expected by increasing the marginal price from 50 cents to 60 cents/1000 gallons (Table 48).

Reduce the Sprinkling Use

The study published by the Northeastern Illinois Planning Commission indicates that about 2 inches of water applied to a lawn over a season are sufficient.[52] Linaweaver, Geyer, and Wolff estimates that in the Eastern US the average actual lawn sprinkling in metered residences amounts to 4.7 inches during a summer.[53]

The first 77 observations in the sample used in this study showed that the average area not covered by buildings was 3445 square feet of which 2400 feet may be assumed to be under grass or shrubs. This would require 5856 gallons over the season or 38.27 gallons/day/residence over the period May through September. If the 4.7 inches are reduced to 2 inches as suggested

---

[52] Northeastern Illinois Planning Commission, The Water Resource in Northeastern Illinois: Planning Its Use, (Chicago: NIPC, 1966), p. 68.

[53] Linaweaver, Geyer, and Wolff, A Study of Residential Water Use, p. 47, table 5.

above the requirement for lawn irrigation would be reduced to 16.25 gallons/day/residence during summer or a reduction of 1.1 MGD in the design city of 50,000 residences. This is comparable to the reduction expected from the alternative of increasing marginal price from 70 to 80 cents/1000 gallons.[54]

Adoption of Odd-Even Rule for Lawn Watering

The demand for water in a municipality on the maximum day is estimated to be about 1.5 times the demand on the average day. In Etobicoke (1967 population 266,458) about half its water was used by residential customers; during the two winter months January and February the average water use was 17.889 MGD and during the two summer months July and August the average water use was 22.32 MGD, an increase of 24.85 per cent. The maximum day use was 39 MGD i.e., an increase of 74.77 per cent over the summer average use.[55]

If one assumes that this peak effect is due to more intensive lawn watering, then one can aim to reduce the peak by half by adopting a municipal by-law, viz., that residences with even door numbers can water their lawns only on even days of the month and residences with odd numbers can water their lawn only on odd days of the month. The reduction in the water use on the maximum day would be 21.38 per cent.

This alternative requires some policing and is also a slight diminution in the standard of service given by most municipal water supply systems, viz., water in virtually unlimited quantities at any time. It is, however, an attractive alternative as a short-term expedient and could well work out over a longer period of time, thus obviating the need to provide more capacity in order to cope with the demand on the maximum day.

Table 50 brings some of the alternatives discussed in this chapter together. Metering and pricing are preferable because they respect the consumers' freedom of choice and they do not

---

[54] The reduction of 1.1 MGD is out of a total that probably exceeds the figure of 6.7255 MGD given in Table 43 above. The latter figure is a conservative estimate of water use on an average summer day.

[55] Unpublished data made available by the Municipal Department of Engineering,

TABLE 50. EXPECTED IMPACT OF SOME POLICY ALTERNATIVES
ON RESIDENTIAL WATER USE

Policy Alternative	Reduction in Water Use Per Cent	Target Variable
Meter and charge a marginal price of 80 cents/1000 gallons	63.25	WUs
Meter and charge a marginal price of 60 cents/1000 gallons	50.10	WUs
Increase marginal price from 40 to 60 cents/ 1000 gallons	35.19	WUs
Increase marginal price from 60 to 80 cents/ 1000 gallons	26.38	WUs
Reduce sprinkling use from an estimated 4.7 to 2.0 inches per season	16.34	WUs
Reduce capacity of flush tank by 50%	16.6	WUs
Adopt "odd-even" by-law on lawn irrigation	21.38	Maximum day use WUm

NOTE: WUs is the water use in gallons/day/dwelling unit (summer period average).
WUm is the water use in gallons/day/dwelling unit (maximum day).

require policing. When the marginal price is already high, or the management wants to reduce water consumption temporarily, the other alternatives should be examined.

V

Summary and conclusions

RESIDENTIAL WATER: PRICING AND INVESTMENT POLICY

Politicians, administrators, planners, and the general public
are often forced to assign priorities to a number of perceived
needs that reflect the preferences of various interest groups in
society. Decisions have at times to be made among various
projects e. g., an increase in the production of material goods,
an increase in outdoor recreation facilities, or in the provision
of better education facilities, and so on. Therefore, the decision-
making process requires vigorous, rational discussion in order
to reach viable decisions.[1] This is particularly important in
situations where the producer (e. g., a large corporation,
government agency) is largely insensitive to market forces as
expressed by consumers' preferences.[2] The rational approach

---

[1] For a complementary viewpoint see Albert Lapawsky, "The Quest for Quality in the
Administration of Resources, " in Natural Resources: Quality and Quantity, eds. S. V.
Ciriacy-Wantrup and J. J. Parsons (Berkeley: University of California Press, 1967),
pp. 162-175.

[2] The implications of weak market forces in resource allocation is discussed by John
Kenneth Galbraith, The New Industrial State (Boston: Houghton Mifflin Co., 1967).

to decision making in natural resource development has been discussed in numerous monographs and manuals dealing with the economics of water resource development projects.[3] However, this literature includes only a small number of studies on municipal water supply economics and policy;[4] the management of demand for municipal and residential water has been almost totally neglected.

Although water is a ubiquitous commodity, the development of a water resource for municipal purposes requires considerable investment which is characterised by discontinuities and which is almost wholly irreversible. Among the major categories of municipal water use, residential water demand imposes extra investment costs due to the widely distributed system of pipes and the peak demands, particularly during the summer. It is important to ascertain to what extent residential consumers are prepared to pay the incremental cost that an addition to the urban water supply system necessitates. This distinction between demand (i.e., a price-quantity schedule), as opposed to 'requirements', is critical to the formulation of rational investment policies for municipal and other essential public services.

Investment in water utilities, according to Bogue, is greater than the combined investment in iron and steel, it is more than twice the investment in gas utilities, and much more than the investment in railroads.[5] Investment in water utilities in the next few decades is likely to be substantial (e.g., $350 - $450 per new resident in North America), partly because of the backlog of dificiencies in facilities, partly because of rapid urban growth. Investment costs per capita, as well as per unit of

[3] For a useful discussion see Ian Burton, "Investment Choices in Public Resource Development," in The Prospect of Change: Proposals for Canada's Future, ed. Abraham Rotstein (Toronto: McGraw-Hill, 1965), pp. 149-173. Other titles published since 1964 include A.V. Kneese and S.C. Smith, eds., Water Research (Baltimore: Johns Hopkins Press 1966); A.V. Kneese and B.T. Bower, Managing Water Quality: Economics, Technology, Institutions (Baltimore: Johns Hopkins Press, 1968); Robert H. Haveman, Water Resource Investment and the Public Interest (Nashville: Vanderbilt University Press, 1965); T.H. Campbell and R.O. Sylvester, eds., Water Resources Management and Public Policy (Seattle: University of Washington Press, 1968).

[4] Besides relevant works referred to so far, see J.S. Bain, R.E. Caves, and J. Margolis, Northern California's Water Industry (Baltimore: Johns Hopkins Press, 1966), particularly chapters 5, 6 and 10.

[5] S.H. Bogue, "Financial Management of a Water Utility," Journal AWWA 60, no. 3 (1968):267.

capacity, are likely to increase due to higher per capita rates of use, more marked peaks in demand, and the need to develop less accessible sources of supply as the present ones prove to be inadequate to meet increasing demand and the tendency for urban areas to 'sprawl'. If new policy alternatives are available, the investment requirements could be reduced; this study shows that the management of demand can reduce investment requirements.

Residential water is priced — albeit imperfectly — and therefore it is possible to test the validity of a widespread prejudice among water utility management, i.e., the way residential water is priced does not affect the level of water use. At the same time the pricing of residential water consumption in general is based on principles which are both theoretically unsound and applied in a haphazard manner. Therefore, the discussion of the price elasticity of residential water is aimed primarily towards the development of more rational water rates schedules. In other words if pricing does affect demand, the incremental cost should be reflected in the water rates schedule. The inclusion of price as a policy variable would also be useful in forecasting and planning future demand for residential water.

In North America the water utility industry may be considered to have outgrown the initial stages when the supply of pure, piped water for essential purposes was clearly justified in terms of improvement in health and the reduction of time spent in carrying water into the house. The decision now is often whether to install further capacity to meet higher per capita demands at peak periods. In the study area, the amalgamation of local governments into multi-tier regional/metropolitan units and the possibilities of subsidies from central funds for 'oversizing' water supply and sewage treatment facilities gives the discussion topical interest since the splitting of management responsibilities may intensify the present tendency to design the water supply system first and price the water afterwards.

The paucity of data on residential water supply/demand may be one reason for the small number of studies in this field. Some of these previous studies used average data and included all municipal water uses and it may be argued that the price-quantity relationship simply reflects the fact that high water using industry avoids municipalities where water rates are high. Other studies use average data from residential water users but do not control for declining block rates which may strengthen the negative correlation between residential water use and marginal price. To eliminate this possibility the data used in this study are the water use of individual households from municipalities that have no reduction in price within the broad range of residential water consumption. Unfortunately this constraint makes it impossible to test empirically the effect of a large allowance of water with the minimum bill.

Some suggestions on the type of data which may be used in subsequent studies may be useful to those interested in further research effort in this field. The review of previous 'models' of residential water use has brought out no consensus of opinion about which factors are most important in 'explaining' the considerable amount of variability in residential water use. The choice of variables seems to depend to a large extent on the availability of data. In the Johns Hopkins study accurate readings of water use for groups of residences were obtained by means of self-recording meters; this is preferable to aggregate data for municipalities. The expense of using self-recording meters limits the size of sample available and the best approach would be to use the self-recording meters as a control for a much larger sample of individual households whose water meter is read on the same day every month. Such a sampling strategy would allow for stratification with respect to 'independent' variables such as income group (or assessed sales value of residence), size of household, age composition of house-hold, size of lawn, marginal charge for water, size of the minimum bill, the amount of water allowed with the minimum bill, climatic conditions, frequency of billing, time interval since the last increase in water rates, the amount of the last increase in the water rates, cultural differences, and so on. If the study is continued over a sufficiently long period of time (say a year or two), some data about the individual households may be obtained by means of a questionnaire. The inclusion of data about the number of days the family spent away on holidays may not be directly useful in policy formulation or the measure-ment of price elasticities but, by reducing the standard error from the regression equation, they make the user of the equation more confident in applying the results of the research. On the other hand, information about the number, type, and frequency of use of water-complementary household appliances is useful in making projections of water use.[6] The monthly reading on water use would reduce some of the errors of measurement that are inherent when the period over which water use is measured varies from household to household.

FINDINGS

It was hypothesized that a structural relationship exists between the average daily residential water use by individual households and a number of 'explanatory' variables. The water use by households was averaged over a year (WUa), the summer period (WUs), and the winter period (WUw). The three measures over- lap to some extent because the billing periods do not always coincide with the thermal seasons. In spite of these errors in the measurement of the dependent variable, the results of this study are encouraging: they are consistent both with a priori expectations and with the recent work by Howe and Linaweaver.[7] Further work along the lines suggested in the previous section should reduce the error of measurement of average residential water use,and thus increase the efficiency of the regression coefficients,as well as reduce the standard error from the equation.

The fitted equations for metered water consumption by house- holds are given below. The initial formulation of the hypothesized relationship included other variables whose effect on residential water use required further investigation; a larger sample stratified with respect to these variables (e.g., lot size, amount of water allowed with the minimum bill, frequency of billing) may provide more reliable information on the structural relationship between residential water use and socio-economic/policy variables. The hypothesized relationship was tested against a sample of house-

[6]See the study funded by the General Electric Co. and designed to analyse the effect of various household appliances on water use. J.S. Anderson and K.S. Watson, "Patterns of Household Usage, " Journal AWWA 59, no. 10 (1967):1228-1237.

[7]Charles W. Howe and F.P. Linaweaver, Jr., "The Impact of Price on Residential Water Demand and Its Relation to System Design and Price Structure, " Water Resources Research 3, no. 1 (1967).

holds from the Toronto Region. The fitted equations for metered and single-unit dwellings are as follows:

$$\log WU_a = 2.78 + 0.56\log V + 0.59\log N_p - 0.93\log P - 0.31\log F \qquad (38)$$
$$(0.13) \qquad (0.08) \qquad (0.22) \qquad (0.14)$$

t-value   4.43**   7.26**   -4.14**   -2.26**

$R = 0.75^{**}$ $S_Y = 0.22$
$R^2 = 0.56$ S.E. $= 0.15$ F-value $= 27.5^{**}$

$$\log WU_s = 3.24 + 0.51\log V + 0.63\log N_p - 1.07\log P - 0.35\log F \qquad (39)$$
$$(0.14) \qquad (0.09) \qquad (0.24) \qquad (0.15)$$

t-value   3.80**   7.29**   -4.50**   -2.40**

$R = 0.74^{**}$ $S_Y = 0.23$
$R^2 = 0.55$ S.E. $= 0.16$ F-value $= 26.45^{**}$

$$\log WU_w = 2.45 + 0.48\log V + 0.62\log N_p - 0.75\log P - 0.24\log F \qquad (40)$$
$$(0.14) \qquad (0.09) \qquad (0.25) \qquad (0.15)$$

t-value   3.37**   6.82**   -3.03**   -1.56

$R = 0.70^{**}$ $S_Y = 0.22$
$R^2 = 0.49$ S.E. $= 0.16$ F-value $= 20.38^{**}$

** Significance at 0.01 level

where WUa is the water use (annual average) in gallons/day/dwelling unit (g/d/du);

WUs is the water use (summer average) in g/d/du;

WUw is the water use (winter average) in g/d/du;

V is the assessed sales value of residence in hundreds of dollars;

Np is the number of persons in dwelling unit;

P is the marginal charge for residential water in cents/1000 gallons;

F is the fixed bill for one billing period in cents; and where logarithms to base 10 are taken for all variables.

The partial regression coefficients are all significantly different from zero and have the expected sign. The fitted equations and the analysis in Chapter III above make it possible to draw some conclusions about the structural relationship of residential water use and their implications for residential water demand management.

The relationship between residential water use and the independent variables is linear in the logarithmic form. This conclusion should be emphasized with respect to price since this variable is the most useful and objective policy instrument available for residential water use management. This form of the equation gives constant elasticities and it may be transformed into a curve convex to the point of origin in the arithmetic form. The present evidence, as well as a priori considerations, suggest that a (small) difference in the marginal charge for water has a greater impact on the level of residential water use when the marginal charges are low.

The fallacy that the demand for residential water is inelastic with respect to price is based on the premise that residential water is a single 'good'. Like land, water is complementary to many activities and it may be said to have a 'composite' demand. Some water uses are likely to be very inelastic (e. g. , water for drinking and washing); other water uses are likely to be price elastic (e. g. , lawn watering).

In fact the price elasticity for water use during the winter period is -0.75 while for the summer period the price elasticity is -1.07.[8] The present evidence suggests that as the range of water uses increases (e. g. , from winter to summer), the impact of price becomes more marked. Therefore, the introduction of a rational water rate schedule (e. g. , marginal pricing or an increasing block schedule) should result in a reduction in peak demands and hence a reduction in the level of investment required.

The analysis of the combined sample of single unit residences (metered) and of groups of townhouses (unmetered) indicates that the regression coefficient with respect to size of households is not significantly different from single-unit dwellings to townhouses. This analysis also throws some light on the impact of metering on water use. The evidence at hand suggests that

[8]This compares well with Howe's reported price elasticity of -0.9 for sprinkling use on an average summer day in the Eastern US. See Charles W. Howe, "Municipal Water Demands, " in Forecasting the Demand for Water, eds. W. R. D. Sewell and B. T. Bower (Ottawa: Queen's Printer, 1968), p. 70

metering per se has little impact on the level of residential water use. Metering, in order to be effective, should be accompanied by high marginal prices. Otherwise the decision to meter residential water does little to curb waste in water use (and hence does not reduce the required investment and running costs) and uses some more resources in the purchase, installation, maintenance, and reading of water meters.

Since metering is not a costless alternative, the introduction of metering should be preceded by an inquiry into the benefits and costs of metering in a specific municipality or group of municipalities. The increase in revenue from water consumption is not a benefit of metering to the community; it is simply a transfer of income within the community, i.e., from householders to the municipal water revenue office. The benefits of metering are twofold: (1) the reduction in the operating costs due to decreased demand for water, and (2) the reduction in the investment required to meet demand that is obviated by metering.

Next an attempt was made to obtain an approximate estimate of the impact of policy alternatives (e.g., an increase in marginal charges) on investment requirements. Data on the cost of municipal water supply, in general, and residential water supply, in particular, were not available and empirical testing was not possible. It is reasonable, however, to expect that, ceteris paribus, lower demands for water at peak periods would require less investment. The proportion by which investment is reduced would depend on the size of the utility, the economies of scale that apply, the interest rate, the source of water, and the pattern of demand (e.g., degree of peakedness). The applications described in Tables 44 and 48 are, therefore, in the nature of worked out examples of possible savings, rather than definite conclusions. On the other hand, the reduction in investment requirements are likely to be conservative.

The community served by the design system is assumed to have the following characteristics: (i) a population of 200,000 people living in 50,000 single-unit dwellings; (ii) a required investment of $150 per resident for residential water supply; (iii) changes in the capacity demand have a proportionate effect on 0.65 of the required investment. These characteristics or assumptions could be changed in other applications.

If $300 per capita is required to build a municipal water supply system, the total required investment is $60.0 million

of which half is ascribed to residential water users. Of this
$30.0 million it is possible to affect 0.65 by reducing the
design capacity. Some examples of policy alternatives and their
expected impact on the level residential water use and the level
of required investment are given below (Table 51).

It is clear that the most substantial reductions occur when
the base marginal charge is low (e.g., unmetered, where the
marginal price = 0). As the marginal price approaches 80
cents/1000 gallons (1967 prices), the difference in marginal
price has a lesser impact on water use and thus on design
capacity. Non-price alternatives are less attractive. For
example, the adoption of the 'odd-even' by-law on lawn irrigation
reduces water use by 21 per cent and it represents a (minor)
diminution in service. The reduction of sprinkling rates from
an estimated 4.7 inches/season to 2.0 inches would reduce water
use on an average summer day by 16 per cent. A national pricing
schedule offers more scope for economies and does not depend
on police action. However, non-price alternatives are useful
for short-term 'crises' and in those municipalities where
marginal charges are already high.

TABLE 51.   THE EFFECT OF DEMAND MANAGEMENT ON
RESIDENTIAL WATER USE AND INVESTMENT REQUIREMENTS

Policy Alternative	Reduction in Water Use (Average Summer day) Per Cent	Reduction in Investment[a]	
		Per Cent	$ Million
Meter, marginal charge = 40 oents/1000 gallons	23	15	4.5
Meter, marginal charge = 60 cents/1000 gallons	50	32.5	9.8
Meter, marginal charge = 80 cents/1000 gallons	63	41	12.3
Charge a marginal price of 60 cents instead of 40 cents/1000 gallons	35	23	6.9
Charge a marginal price of 80 cents instead of 60 cents/1000 gallons	26	17	5.1

[a] Total investment ascribed to residual water use is $30 million.

190

The most preferred type of pricing schedule is an increasing block schedule because small consumers (who are least likely to contribute to the peak demand) are not penalized. Even if small consumers of residential water are subsidized, this would reflect the 'social good' aspect of the provision of piped, treated water to urban dwellers. On the other hand, those consumers who are able to pay for more water (and thus necessitate an increase in capacity) would be provided with this water and asked to pay the marginal cost. In addition, the pricing schedule with increasing blocks is simple to understand and administer; it would ensure the recovery of expenditures through revenues; it would have the same purpose as summer charges without complicating the operations of the water revenue branch; it would tend to allocate resources to water supply in such a way that both over-and-under-building are avoided. In short the increasing price schedule is effective, fair, simple, and efficient.

SOME TENTATIVE RECOMMENDATIONS

The stance taken in this study reflects a conviction that theoretical analysis and practical application complement each other; one derives meaning and relevance from the other. It would be useful to indicate briefly some of the problems that require attention and to make some tentative recommendations for consideration at the initial stage of implementation.

1. Policy variables, in particular metering and the level of the marginal commodity charges, should be taken into account in forecasting and designing capacity for residential water supply. The assumption that residential water use is inelastic to price at the margin is demonstrably untenable. The evidence from the present and other independent studies in fact suggests that the price elasticity for water use on an average day in summer in Eastern North America is close to unity.
2. Unified management of municipal water utilities gives fuller scope to the articulation of policy objectives and instruments than the fragmentation of management functions among different levels of government. Marginal pricing or metering have little

appeal where the benefits of such policies are in part passed on to other jurisdictions.

3. Where it can be determined that the average cost of additions to capacity are higher than present average costs, the schedule of water rates should reflect this trend. The minimum bill, under such circumstances, should not be increased when water rate schedules are revised and the amount of water allowed with the minimum bill should not exceed (about) 2000 gallons/month/dwelling unit. Eventually an increasing block schedule should be established in all municipalities where the water supply is characterized by increasing average costs with respect to new additions to capacity.

4. Further studies are needed to investigate the impact of introducing metering in those municipalities where residential water use is not metered and of metering townhouses individually rather than collectively. Such benefit-cost analyses should include consideration of investment reductions or postponements as a result of metering. It is also essential to take into account the fact that metering involves a marginal price and the setting of this price is part of the general decision to introduce metering.

5. Further study is also required to determine the impact of price increase over time in the Toronto Region and to explore the possible effects of non-price alternatives such as smaller household plumbing fixtures and the odd-even rule for lawn watering.

The tentative findings and suggestions would hopefully stimulate further discussion and research into this resource-allocation problem that is likely to demand increasing attention in an age of rapid urban growth. In resource policy formulation and planning one has to steer a course between the twin dangers of 'paralysis by analysis' and of 'decision by instinct.' The water utility industry has achieved an enviable reputation of high professional standards by its insistence on high quality water and service to its customers. It is in large measure a result of successful endeavour in the past that residential water has ceased to be a social good and has become an economic good. The formulation of an investment policy,based on consumer demand rather than on the 'requirements' approach,would ensure that scarce public funds are allocated to projects that reflect the preferences and priorities of the consuming public; in turn this management strategy would ensure that scarce public funds are allocated to projects that reflect the preferences and priorities of the consuming public; in turn this management strategy would ensure public support for this essential service in the urban age. Finally, the range and impact of alternatives available to management is likely to vary over space and time; neither hallowed 'rule of thumb' nor sweeping generalities can substitute for detailed study of this basic resource-use problem in its regional setting.

Bibliography

Materials relating to Ontario and the study area are listed in a separate section at the end of this bibliography.

American Water Works Association is abbreviated to AWWA.

ACKERMAN, E. A. AND G.O.G. LOF. Technology in American Water Development. Baltimore: Johns Hopkins Press, 1959.

ACTON, F. S. Analysis of Straight-Line Data. New York: Wiley, 1959.

AFIFI, H.H.H. "Economic Evaluation of Water Supply Pricing in Illinois." Journal AWWA 61, no. 1 (1969):41-48.

AMERICAN CITY MAGAZINE. Modern Water Rates. New York: Buttenheim, 1965.

AMERICAN SCHOOL OF CLASSICAL STUDIES IN ATHENS. Waterworks in the Athenean Agora. Princeton, NJ: 1968.

AMERICAN WATER RESOURCES ASSOCIATION. Proceedings, Fourth American Water Resources Conference, New York, 1968. Urbana, Ill.: AWRA, 1968.

--------. Hydata. An international review of the contents of periodicals in the field of water resources. Monthly. January 1965-   Urbana, Ill.: AWRA.

AWWA COMMITTEE REPORT. "Determination of Water Rates Schedules." Journal AWWA 46, no. 3 (1954):187-219. (Republished as Water Rates Manual. New York: AWWA, 1957.)

--------. "Basic Principles of a National Water Resources Policy." Journal AWWA 49, no. 7 (1957):825-833.

--------. "Trends in Air-Conditioning Use and Regulation." Journal AWWA 50, no. 1 (1958):75-96.

--------. "Trends in Air-Conditioning Regulation." Journal AWWA 57, no. 11 (1965):1456-1471.

AWWA PANEL DISCUSSION. "What is Good Water Service and How Should it be Paid For." Journal AWWA 55, no. 1 (1963):4-26.

AWWA STAFF REPORT. "A Survey of Operating Data for Waterworks in 1955." Journal AWWA 49, no. 5 (1957):553-696.

--------. "The Water Utility Industry in the United States." Journal AWWA 58, no. 7 (1966):767-785.

ANDERSON, J. S. AND K. S. WATSON. "Patterns of Household Usage." Journal AWWA 59, no. 10 (1967):1228-1237.

ARGO, T. W. "Factors to Consider in Developing Water Rate Structures." In Canadian Municipal Utilities: Waterworks Manual and Directory, 1962. Annual. Toronto: Monetary Times Publications, 1962.

AZPURUA, P. P.; A. S. EDUARDO; AND P. F. M. RUIZ. "New Water Rates for Caracas." Journal AWWA 60, no. 7 (1968):774-780.

BABBITT, H. E.; J. J. DOLAND; AND J. L. CLEASBY. Water Supply Engineering. 6th ed. New York: McGraw-Hill, 1962.

BAIN, J. S.; R. E. CAVES; AND J. MARGOLIS. Northern California's Water Industry. Baltimore: Johns Hopkins Press, 1966.

BAKER, M. N. The Quest for Pure Water: The History of Water Purification from the Earliest Records to the Twentieth Century. New York: AWWA, 1949.

BARROWS, HARLAN A. "Geography as Human Ecology." Annals Association of American Geographers 13, no. 1 (1923):1-14.

BAXTER, S. S. "Principles of Rate Making for Publicly Owned Utilities." Journal AWWA 52, no. 10 (1960):225-238.

--------. "Water Supply: Economics, Technology and Policy." Journal AWWA 55, no. 9 (1963):1225-1228.

BECKWITH, H. E. "Economics of Leak Surveys." Journal AWWA 56, no. 5 (1964):575-578.

BERRY, A. E. "A Tribute to Norman Howard Joseph." Journal AWWA 56, no. 10 (1964).

--------. "Environmental Pollution and Its Control in Canada: A Historical Perspective." In Pollution and Our Environment, Background Paper A-1, Canadian Council of Resource Ministers Conference, Montreal, 1966. Vol. 1. Ottawa: Queen's Printer, 1967.

BIRD, P. A. AND C. I. JACKSON. "Economic Methods of Charging for Water." Journal British Waterworks Association 48, no. 419 (1966):614-628.

--------. "Economic Charges for Water." In Essays in the Theory and Practice of Pricing. London: Institute of Economic Affairs, 1967.

BLAKE, N. M. Water for the Cities: A History of the Urban Water Supply Problem in the U.S. Syracuse: Syracuse University Press, 1956.

BOGUE, S. H. "Trends in Water Use." Journal AWWA 55, no. 5 (1963):548-554.

--------. "Financial Management of a Water Utility." Journal AWWA 60, no. 3 (1968): 267-272.

BOJESSON, E. K. G. AND C. M. BOBEDA. "New Concept in Water Service for Developing Countries." Journal AWWA 56, no. 7 (1964):853-862.

BONEM, G. W. "On the Marginal Cost Pricing of Municipal Water." Water Resources Research 41, no. 1 (1968):191-193.

BOULDING, K. E. "The Economist and Engineer: Economic Dynamics of Water Resource Development." In Economics and Public Policy in Water Resource Development, edited by S. C. Smith and E. N. Castle. Ames: Iowa State University Press, 1964.

BROCK, D. A. "Multiple Regression Analysis of Maximum Day Water Consumption of Dallas, Texas." Journal AWWA 50, no. 10 (1958):1391-1394.

BROCK, R. P. "Concerning Services and Rates." Journal AWWA 61, no. 8 (1969):413.

BRUCE, J. P. AND D. E. L. MAASLAND. Water Resources Research in Canada. Science Secretariat, Privy Council Office Special Study No. 5. Ottawa: Queen's Printer, 1968.

BURDICK, C. E. "Storage on the Distribution System." Water and Sewage Works 99 (1952), Reference and Data Section, R-35 ff.

BURTON, IAN. "Investment Choices in Public Resource Development." In The Prospect for Change: Proposals for Canada's Future, edited by Abraham Rotstein. Toronto: McGraw-Hill, 1965.

BURTON, IAN AND ROBERT W. KATES. "Canadian Resources and American Requirements." Canadian Journal of Economics and Political Science 30, no. 2 (1964):265-269.

BUTLER, IAN. "The Role of Water-Tanks and Reservoirs." In Canadian Municipal Utilities: Waterworks Manual and Directory, 1964-65. Annual. Toronto: Monetary Times Publications, 1965.

CAMPBELL, T. H. AND R. O. SYLVESTER, eds. Water Resources Management and Public Policy. Seattle: University of Washington Press, 1968.

CANADA. Science Council of Canada. A Major Program of Water Resources Research in Canada. Report No. 13. Ottawa: Queen's Printer, 1968.

CANADIAN COUNCIL OF RESOURCE MINISTERS. Pollution and Our Environment. Conference Background Papers, prepared for the National Conference on Pollution and Our Environment, Montreal, 1966. Ottawa: Queen's Printer, 1967.

--------. The Administration of Water Resources in Canada. Montreal, 1968.

197

Canadian Municipal Utilities: Waterworks Manual and Directory. Toronto: Monetary
Times Publications, published annually.

CASS-BEGGS, D. "Water as a Basic Resource." In Water: Process and Method in
Canadian Geography, edited by J. G. Nelson and M. J. Chambers. Toronto:
Methuen, 1969.

CAULFIELD, H. P. "Municipal Water in Federal Programs." In Water: Development,
Utilization, Conservation, edited by R. K. McNickle. Boulder: University of
Colorado Press, 1964.

CHORLEY, R. J., ed. Water, Earth and Man: A Synthesis of Hydrology, Geomorphology
and Socio-Economic Geography. London: Methuen, 1969.

CIRIACY-WANTRUP, S. V. "Water Policy and Economic Optimizing: Some Conceptual
Problems in Water Research." American Economic Review, Papers and Proceedings 57 (1967):179-189.

CIRIACY-WANTRUP, S. V. AND J. J. PARSONS, eds. Natural Resources: Quality and
Quantity. Berkeley: University of California Press, 1967.

CLARK, J. W. AND W. VIESSMAN, Jr. Water Supply and Pollution Control. Scranton,
Pa.: International Textbook Co., 1965.

CLARK, R. H.; A. K. WATT; AND J. P. BRUCE. "Basic Data Requirements for Water
Management." In Resources for Tomorrow Conference, Background Papers,
Montreal, 1961. Vol. 1. Ottawa: Queen's Printer, 1961.

COBB, E. B. AND W. H. BROWN. "Remote Sensing of Water Meters." Journal AWWA
59, no. 11 (1967):1440-8.

CORNELL, H. "Comparative Water System Plan." Journal AWWA 60, no. 2 (1968):125-
128.

CRAINE, L. E. "Water Management and Urban Planning." American Journal of Public
Health 51, no. 3 (1961):427-433.

CREW, M. A. "Pricing for Efficiency in Electricity Supply." In Essays in the Theory
and Practice of Pricing. London: Institute of Economic Affairs, 1967.

DAVIS, R. K. Range of Choice in Water Management: A Study of Dissolved Oxygen in
the Potomac Estuary. Baltimore: Johns Hopkins Press, 1968.

DeHAVEN, J. C. "Water Supply Economics, Technology and Policy." Journal AWWA
55, no. 5 (1963):539-47.

DIETERICH, BERND H. AND J. M. HENDERSON. Urban Water Supply Conditions and
Needs in Seventy-Five Developing Countries. World Health Organization Public
Health Paper No. 23. Geneva: WHO, 1963.

DRAPER, N. R. AND H. SMITH. Applied Regression Analysis. London: Wiley, 1966.

DUNN, D. F. AND T. E. LARSON. "Relationship of Domestic Water Use to Assessed
Valuation with Selected Demographic and Sociometric Variables." Journal AWWA
55, no. 4 (1963):441-449.

198

EATON, E. D. "Water Resources Planning Strategy." In Water Resources Planning, Conference, New England Council of Water Center Directors, Boston, May 1968.

ELDER, C. C. "Determining Future Water Requirements." Journal AWWA 43, no. 2 (1951):124-135.

EZEKIEL, M. AND K. A. FOX. Methods of Correlation and Regression Analysis: Linear and Curvilinear. New York: Wiley, 1959.

FAIR, G. M. AND J. C. GEYER. Water Supply and Waste-Water Disposal. 6th print. New York: Wiley, 1965.

FOUST, R. J. "N.E. Water Crisis and Its Solution." Journal AWWA 58, no. 1 (1966): 3-7.

FEDERAL COUNCIL FOR SCIENCE AND TECHNOLOGY, COMMITTEE ON WATER RESOURCES RESEARCH. A Ten-Year Program of Federal Water Resources Research. Washington, DC: GPO, 1966.

FEDERATION OF SCIENTIFIC AND TECHNICAL ASSOCIATIONS, WATER STUDY GROUP. Water for Tomorrow: An International Survey. Milan: Mondadori, 1969.

FINCH, L. S. "Phoney Hydromania: The Real Danger to Our Water Supplies." Water and Wastes Engineering 3, no. 1 (1966):36-38.

FLACK, J. E. "Meeting Future Water Requirements Through Reallocation." Journal AWWA 59, no. 11 (1967):1340-1349.

FOURT, L. "Forecasting the Residential Demand for Water." Seminar Paper, Agricultural Economics, University of Chicago, February 1968. Mimeographed.

FOX, IRVING K. "We Can Solve Our Water Problems." Water Resources Research 2, no. 4 (1966):617-123.

FOX, KARL A. Intermediate Economic Statistics. New York: Wiley, 1968.

FRANCIS, J. L. "Designing Residential Water Services." Journal AWWA 62, no. 2 (1970):85-90.

FRANKEL, R. J. "Water Quality Management: Engineering-Economic Factors in Municipal Waste Disposal." Water Resources Research 1, no. 2 (1965):173-186.

FRONTINUS. De Aquis Urbis Romae. Translated by Clemens Hershel. Boston: Dana Estes, 1899.

GALBRAITH, JOHN KENNETH. The New Industrial State. Boston: Houghton Mifflin, 1967.

GOPÁLAKRISHNAN, C. "Water Transfer and Economic Development." Water Resources Bulletin 4, no. 2 (1968):45-50.

GOTTLIEB, M. "Urban Domestic Demand for Water: A Kansas Case Study." Land Economics 39, no. 2 (1963):204-210.

GRAVA, SIGURD. Urban Planning Aspects of Water Pollution Control. New York: Columbia University Press, 1969.

199

GRAESER, H. J. "The Water Industry and Local Government." Journal AWWA 60, no. 1 (1968):1-5.

GRAYSON, L. W. "Water Supply: America's Greatest Challenge." Journal AWWA 52, no. 1 (1960):1-5.

GRUNFELD, Y. "The Interpretation of Cross-Section Estimates in a Dynamic Model." Econometrica 29 (1961):397-405.

HAAR, C. M. "Water Distribution Systems Interlocked for Mutual Gain." Public Works 98, no. 11 (1967):100-101.

HANKE, S. T. AND J. E. FLACK. "Effects of Metering Urban Water." Journal AWWA 60, no. 12 (1968):1359-65.

HANSEN, A. E. "Pressure Records." Journal AWWA 47, no. 2 (1955):159-162.

HARDENBERGH, W. A. AND E. R. RODIE. Water Supply and Waste Disposal. Scranton, Pa.: International Textbook Co., 1961.

HART (REES), JUDITH. "Demand for Water by Manufacturing Industry in South East England." M. Phil. thesis, University of London, 1968.

HATCHER, M. P. "Basis for Rates." Journal AWWA 57, no. 3 (1965):273-278.

HAVEMAN, R. H. Water Resources Investment and the Public Interest. Nashville: Vanderbilt University Press, 1965.

HEADLEY, J. C. "The Relation of Family Income and Use of Water for Residential and Commercial Purposes in the San Francisco-Oakland Metropolitan Area." Land Economics 39, no. 4 (1963):441-449.

HEANEY, J. P. AND R. S. GEMMELL. "Production Cost Factor in Rate Making." Journal AWWA 61, no. 2 (1969):102-106.

HEGGIE, G. D. "Effects of Sprinkling Restrictions." Journal AWWA 49, no. 3 (1957): 267-275.

HENDERSON, A. D., et al. "The Lawn Sprinkling Load." Journal AWWA 48, no. 4 (1956):361-377.

HERSCHEL, CLEMENS. The Two Books on the Water Supply of the City of Rome of Sextus Julius Frontius .... A.D. 97. Boston: Dana Estes and Co., 1899.

HINES, LAWRENCE G. "The Role of Price in the Allocation of Water Resources." Proceedings of the American Society of Civil Engineers (January 1960):15-28.

--------. "The Long-Run Cost Function of Water Production for Selected Wisconsin Communities." Land Economics 45, no. 1 (1969):133-140.

HIRSHLEIFER, JACK; JAMES C. DeHAVEN; AND JEROME W. MILLIMAN. Water Supply: Economics, Technology and Policy. Chicago: University of Chicago Press, 1960.

HIRSHLEIFER, JACK AND JEROME W. MILLIMAN. "Urban Water Supply: A Second Look." American Economic Review, Papers and Proceedings 57 (1967):169-178.

HOEL, P. G. Introduction to Mathematical Statistics. 3rd ed. New York: Wiley, 1963.

HOWE, CHARLES W. "Water Resources and Regional Economic Growth in the U.S., 1950-1960." Southern Economic Journal 34, no. 4 (1968):477-489.

--------. "Water Pricing in Residential Area." Journal AWWA 60, no. 5 (1968):497-501.

--------. "Municipal Water Demand." In Forecasting the Demand for Water, edited by W.R.D. Sewell and B. T. Bower. Ottawa: Queen's Printer, 1968.

HOWE, CHARLES W. AND F. P. LINAWEAVER, Jr. "The Impact of Price on Residential Water Demand and Its Relation to System Design and Price Structure." Water Resources Research 3, no. 1 (1967):13-32.

HOWELLS, D. H. "...nor any drop to drink...unless we conserve municipal water supplies." Water and Wastes Engineering 3, no. 10 (1966):40-43.

HOWSON, L. R. "The Distribution System." Water and Sewage Works 99 (1952), Reference and Data Section, R-19 ff.

--------. "Fifty Years' Experience with Water Rates and Revenues." Journal AWWA 51, no. 6 (1959):693-700.

--------. "Revenues, Rates and Advance Planning." Journal AWWA 52, no. 2 (1960): 153-161.

--------. "Review of Rate Making Theories." Journal AWWA 58, no. 7 (1966):849-855.

HUDSON, W. D. "Reduction of Unaccounted for Water." Journal AWWA 56, no. 2 (1964):143-148.

--------. "Field Testing of Large Meters." Journal AWWA 58, no. 7 (1966):867-873.

HUMMEL, D. "Trends in Metropolitan Water Development." In Water: Development, Utilization, Conservation, edited by R. K. McNickle. Boulder: University of Colorado Press, 1964.

JACKSON, C. I. AND P. A. BIRD. "Water Supply--The Transformation of an Industry." The Three Banks Review (March 1967):23-25.

JOHNSTON, J. Econometric Methods. International Students Edition. New York: McGraw-Hill, 1963.

JUDY, R. W. "Municipal Water Demand: A Critique." In Forecasting the Demand for Water, edited by W.R.D. Sewell and B. T. Bower. Ottawa: Queen's Printer, 1968.

KAY, G. H.; A. R. TOWNSEND; AND K. F. LETHBRIDGE. "Water Pollution Control and Regional Planning--The Grand River Watershed." In Pollution and Our Environment, Background Paper B 17-1-13, Canadian Council of Resource Ministers Conference, Montreal, 1966. Ottawa: Queen's Printer, 1967.

KELLER, C. W. "Design of Water Rates." Journal AWWA 58, no. 3 (1966):293-299.

KEMPTON, L. D. "Air Conditioning Brings Water Problems." Public Works 87, no. 9 (1956):132-134.

KENDALL, M. G. "Regression, Structure, and Functional Relationship, Part 1." Biometrika 38 (1951):11-25.

KNEESE, A. V. Economic and Related Problems in Contemporary Water Resources Management. Resources for the Future Reprint No. 55. Washington, DC: RFF, November 1965.

--------. Economics and the Quality of the Environment: Some Empirical Experiences. Resources for the Future Reprint No. 71. Washington, DC: RFF, 1968.

KNEESE, A. V. AND B. T. BOWER. Managing Water Quality: Economics, Technology and Institutions. Baltimore: Johns Hopkins Press, 1968.

KNEESE, A. V. AND S. C. SMITH, eds. Water Research. Baltimore: Johns Hopkins Press, 1966.

KUH, E. "The Validity of Cross Sectionally Estimated Behaviour Equations in Time Series Applications." Econometrica 27 (1959):197-212.

KUIPER, E. Water Resources Development: Planning, Engineering, Economics. London: Butterworths, 1965.

LANGBEIN, W. B. AND W. G. HOYT. Water Facts for the Nation's Future. New York: Ronald, 1959.

LARSON, B. O. AND H. E. HUDSON. "Residential Water Use and Family Income." Journal AWWA 43, no. 8 (1951):603-610.

LASKIN, BORA. "Jurisdictional Framework for Water Management." In Resources for Tomorrow Conference, Background Papers, Montreal 1961. Vol. 1. Ottawa: Queen's Printer, 1961.

LEARNED, A. P. "Determination of Municipal Water Rates." Journal AWWA 49, no. 2 (1957):165-73.

LEE, TERENCE R. Residential Water Demand and Economic Development. University of Toronto Department of Geography Research Publications No. 2. Toronto: University of Toronto Press, 1969.

LEPAWSKY, A. "The Quest for Quality in the Administration of Resources." In Natural Resources: Quality and Quantity, edited by S. V. Ciriacy-Wantrup and J. J. Parsons. Berkeley: University of California Press, 1967.

LINAWEAVER, F. P., Jr. AND J. C. GEYER. "Use of Peak Demands in Determination of Residential Rates." Journal AWWA 56, no. 4 (1964):403-410.

LINAWEAVER, F. P., Jr. AND C. SCOTT CLARK. "Costs of Water Transmission." Journal AWWA 56, no. 12 (1964):1549-1560.

LINAWEAVER, F. P., Jr.; J. C. GEYER; AND J. B. WOLFF. A Study of Residential Water Use. Washington, DC: GPO, 1967.

LINSLEY, R. K. AND J. B. FRANZINI. Water Resources Engineering. New York: McGraw-Hill, 1964.

LLOYD, D. F. "Cost Comparison of Unmetered and Metered Systems at Idaho Falls." Journal AWWA 52, no. 4 (1960):433-436.

LOF, G.O.G. AND C. H. HARDISON. "Storage Requirements of Water in the U.S." Water Resources Research 2, no. 3 (1966):323-354.

LOGAN, JOHN. "The International Municipal Water Supply Program: A Health and Economic Appraisal." American Journal of Tropical Medicine and Hygiene 9, no. 5 (1960):469-476.

McDONALD, N. G. AND R. A. G. SIMMONS. "Some Economic Factors in the Selection of Water Storage Facilities." In Canadian Municipal Utilities: Waterworks Manual and Directory, 1964-5. Annual. Toronto: Monetary Times Publications, 1965.

McDOUGALL, H. "How do you Assess Water Rates?" In Canadian Municipal Utilities: Waterworks Manual and Directory, 1963. Annual. Toronto: Monetary Times Publications, 1963.

McGANN, P. W. "Technological Progress and Minerals." In Natural Resources and Economic Growth, edited by J. J. Spengler. Washington, DC: Resources for the Future, 1961.

McGAUHEY, P. H. Engineering Management of Water Quality. New York: McGraw-Hill, 1968.

McGINNIS, L. R. "Waterworks Standardization in Distribution and Metering Equipment Impracticable." In Canadian Municipal Utilities: Waterworks Manual and Directory, 1963. Annual. Toronto: Monetary Times Publications, 1963.

McKEE, J. E. AND H. W. WOLF. Water Quality Criteria. 2nd ed. Sacramento, California: State Water Quality Control Board, 1963.

METCALF, L. "Effect of Water Rates and Growth in Population Upon Per Capita Consumption." Journal AWWA 15, no. 1 (1926):1-21.

METZLER, D. F. AND H. B. RUSSELMAN. "Wastewater Reclamation as a Water Resource." Journal AWWA 60, no. 1 (1968):95-102.

MILLER, M. "The Developmental Framework for Resource Policy and Its Jurisdictional-Administrative Implications." Canadian Public Administration Journal 5, no. 2 (1962):133-155.

MILLIMAN, JEROME W. "Economic Aspects of Public Water Utility Construction." Journal AWWA 50, no. 7 (1958):839-845.

--------. "Policy Horizons for Future Urban Water Supply." Land Economics 39, no. 2 (1963):109-132.

--------. "New Price Policies for Municipal Water Service." Journal AWWA 56, no. 2 (1964):125-131. (See comment by E. D. Bovet and reply by Milliman in Journal AWWA 56, no. 11 [1964]:1495-1498; and comment by L. E. Ayres and reply by Milliman in Journal AWWA 56, no. 12 [1964]:1595-1599.)

MOSER, C. A. Survey Methods in Social Investigation. London: Heinemann, 1958.

MOSS, FRANK. The Water Crisis. New York: F. A. Praeger, 1967.

MURDOCK, J. H. "75 Years of Too Cheap Water - 9 Years Later." Journal AWWA 57, no. 8 (1965):943-947.

NATIONAL ACADEMY OF SCIENCE - NATIONAL RESEARCH COUNCIL, COMMITTEE ON WATER. Alternatives in Water Management. Washington, DC: NAS-NRC, 1966.

NORTHEASTERN ILLINOIS PLANNING COMMISSION. The Water Resource in Northeastern Illinois: Planning Its Use. Prepared by J. R. Sheaffer and A. J. Zeizel, et al. Chicago: NIPC, 1966.

NELSON, J. G. AND M. J. CHAMBERS. Water: Process and Method in Canadian Geography. Toronto: Methuen, 1969.

OKUN, D. "Alternatives in Water Supply." Journal AWWA 61, no. 5 (1969):215-221 and "Discussion," 221-224.

ORLOB, G. T. AND M. R. LINDORF. "Cost of Water Treatment in California." Journal AWWA 50, no. 1 (1958):45-55.

PAN AMERICAN SANITARY BUREAU. Facts on Health Problems: Health in Relation to Social Improvement in the Americas. Washington, DC: Pan American Health Organization, July 1961.

PATTERSON, W. L. "Demand Rates for Water Service." Journal AWWA 53, no. 10 (1961):1261-1268.

--------. "Practical Water Rate Determination." Journal AWWA 54, no. 8 (1962):904-912.

--------. "Comparison of Elements Affecting Rates in Water and Other Utilities." Journal AWWA 57, no. 5 (1965):554-560.

PAYNE, B. "Shrinking Nonrevenue Water." Water and Wastes Engineering 3, no. 8 (1966):28-30.

PEARSON, E. S. AND H. O. HARTLEY, eds. Biometrika Tables for Statisticians, Volume 1. Cambridge: Cambridge University Press, 1966.

PORGES, R. "Factors Affecting Per Capita Water Consumption." Water and Sewage Works 104, no. 5 (1957):199-204.

PUBLIC WORKS SURVEY. "Present and Future Estimates of Water Consumption." Public Works 87, no. 13 (1956):73 ff.

QUAKENBUSH, T. H. AND J. T. PHELAN. "Irrigation Water Requirements of Lawns." Journal, Irrigation and Drainage Division, American Society of Civil Engineers 91, IR2 (June 1965):11-20.

RENSHAW, E. F. "Value of an Acre-Foot of Water." Journal AWWA 50, no. 3 (1958): 303-309.

RUSSELL, C. S. The Definition and Measurement of Drought Losses: The Northeast Drought of 1962-66. Resources for the Future Reprint No. 77. Washington, DC: RFF, 1969.

RYNDERS, A. "Demand Rates and Metering Equipment at Milwaukee." Journal AWWA 52, no. 10 (1960):1239-1243.

SCARATO, R. F. "Time-Capacity Expansion of Urban Water Systems." Water Resources Research 5, no. 5 (1969):929-936.

SCHMID, G. C. "Peak Demand Storage." Journal AWWA 48, no. 4 (1956):378-385.

SEASTONE, D. A. AND L. M. HARTMAN. "Alternative Institutions for Water Transfers: The Experience in Colorado and New Mexico." Land Economics 39, no. 1 (1963):31-44.

SEARCY, P. E. AND T. de S. FURMAN. "Water Consumption by Institutions." Journal AWWA 53, no. 9 (1961):1111-1119.

SEIDEL, H. F. AND E. R. BAUMANN. "A Statistical Analysis of Water Works Data for 1955." Journal AWWA 49, no. 12 (1957):1531-1566.

SEIDEL, H. F. AND J. L. CLEASBY. "A Statistical Analysis of Water Works Data for 1960." Journal AWWA 58, no. 12 (1966):1507-1527.

SEMPLE, ELLEN C. "Domestic and Municipal Waterworks in Ancient Mediterranean Lands." Geographical Review, 21, no. 4 (1931):466-474.

SEWELL, W.R.D. "Pipedream or Practical Possibility?" In Water: Process and Method in Canadian Geography, edited by J. G. Nelson and M. J. Chambers. Toronto: Methuen, 1969.

SEWELL, W.R.D.; B. T. BOWER; et al. Forecasting the Demand for Water. Ottawa: Queen's Printer, 1968.

SEWELL, W.R.D. AND IAN BURTON. "Recent Innovations in Resource Development Policy in Canada." Canadian Geographer 11, no. 4 (1967):327-340.

SEWELL, W.R.D.; R. W. JUDY; AND L. OUELLETT. Water Management Research: Social Science Priorities. Ottawa: Queen's Printer, 1969.

SHEAFFER, J. R., et al. Metropolitan Water Resource Management as an Emerging Specialized Technical Area: A State of the Art and Literature Review. Chicago: Center for Urban Studies, University of Chicago, 1969.

SHIPMAN, HAROLD R. "Water Rates Structures in Latin America." Journal AWWA 59, no. 1 (1967):3-12. (See comment by P. P. Azpurua and reply by Shipman in Journal AWWA 60, no. 6 [1968]:743-748.)

--------. "Water Supply Problems in Developing Countries." Journal AWWA 59, no. 7 (1967):767-772.

SIMMONS, J. G. "Economic Significance of Unaccounted for Water." Journal AWWA 58, no. 6 (1966):639-641.

SMALL, F. L. "Planning a Municipal Water System." In Canadian Municipal Utilities: Waterworks Manual and Directory 1963. Annual. Toronto: Monetary Times Publications, 1963.

SMITH, R. L. "Total Management of Water Resources." Journal AWWA 59, no. 11 (1967):1335-1339.

SMITH, S. AND E. N. CASTLE, eds. Economics and Public Policy in Water Resource Development. Ames: Iowa State University Press, 1964.

SNEDECOR, G. W. AND W. G. COCHRAN. Statistical Methods. 6th ed. Ames: Iowa State University Press, 1967.

STANTON PIPES (LTD.). "1967 5th Annual Survey of Municipal Water Rates of Ontario."
Water Works Digest 9, no. 1 (1967). Hamilton, Ontario: Stanton Pipes (Canada)
Ltd.

STRAND, J. A. "Method for Estimation of Future Distribution System Demand."
Journal AWWA 58, no. 5 (1966):521-525.

TANGHE, E. F. "Meeting Milwaukee Peak Demands." Journal AWWA 47, no. 1 (1955):
60-62.

TINBERGEN, JAN. Economic Policy: Principles and Design. 4th rev. print. Amsterdam: North-Holland Publishing Co., 1967.

TURNEARE, F. E. AND II. L. RUSSELL. Public Water Supplies: Requirements,
Resources and the Construction of Works. 4th ed. New York: Wiley, 1940.

TURNOVSKY, S. J. "The Demand for Water: Some Empirical Evidence on Consumers'
Response to Commodity Uncertain in Supply." Water Resources Research 5, no. 2
(1969):350-361.

UNITED NATIONS. Department of Economic and Social Affairs, Resources and Transport Division. Water Desalination in Developing Countries. New York: UN, 1964.

UNITED STATES. Congress. Senate Select Committee on National Water Resources.
Future Water Requirements for Municipal Use. 86th Cong., 2nd Sess., Comm.
Print No. 7. Washington, DC: GPO, 1960.

--------. Congress. Senate Select Committee on National Water Resources. Water
Resources in the United States. 86th Cong., 2nd Sess., Comm. Print No. 31.
Washington, DC: GPO, 1960.

--------. Division of Water Supply and Pollution Control. Basic Data Branch. Municipal Water Facilities in Communities of 25,000 Population and Over. Washington,
DC: GPO, 1962.

--------. Division of Water Supply and Pollution Control. Basic Data Branch.
Statistical Summary of Municipal Water Facilities in the US. Washington, DC:
GPO, 1965.

--------. Federal Housing Administration. Minimum Design Standards for Community
Water Supply Systems. Washington, DC: GPO, 1968.

--------. Inter-Agency Committee on Water Resources. Annotated Bibliography on
Hydrology (1951-54) and Sedimentation (1950-54), United States and Canada.
Washington, DC: GPO, 1955.

--------. Inter-Agency Committee on Water Resources. Annotated Bibliography...
1959-62. (A bibliography for the period 1955-58 was published in US Geological
Survey Water Supply Paper No. 1546.)

--------. Task Force on Environmental Health and Related Problems. A Strategy
for a Livable Environment. Washington, DC: GPO, 1967.

VANCE, J. A.; K. E. SYMONS; AND D. A. McTAVISH. "The Diverse Effects of Water
Pollution on the Economy: Domestic and Municipal Water Use." In Pollution and
Our Environment, Background Paper A 4-1-5, Canadian Council of Resource
Ministers Conference, Montreal 1966. Vol. 1. Ottawa: Queen's Printer, 1967.

VEATCH, N. T. "Use of Water in Arid and Semi-Arid Cities." Journal AWWA 23, no. 7 (1931):937-954.

WARFORD, J. J. "Water Requirements: The Investment Decision in the Water Supply Industry." The Manchester School 34, no. 1 (1966):87-112.

WATERFIELD, D. Continental Waterboy. Toronto: Clarke Irwin, 1970.

WELLISCH, H. Water Resources Development, 1950-65. An International Bibliography. Jerusalem: Israel Program for Scientific Translations, 1967.

WHITE, GILBERT F. Strategies of American Water Management. Ann Arbor: University of Michigan Press, 1969.

--------. Preface to Alternatives in Water Management. National Academy of Sciences--National Research Council, Committee on Water. Washington, DC: NAS-NRC, 1966.

--------. ed. Water, Health, and Society. Selected Papers by Abel Wolman. Bloomington, Ind.: Indiana University Press, 1969.

WHITE, GILBERT F.; D. J. BRADLEY; AND ANNE V. WHITE. Drawers of Water: Domestic Water Use in East Africa. In press.

WILLIAMS, E. J. Regression Analysis. New York: Wiley, 1959.

WINNIPEG. Metropolitan Corporation of Greater Winnipeg, Waterworks and Waste Disposal Division. Report of the Water Supply Study: Water Requirements to 2020. Winnipeg: 1967.

WOLFF, J. B. "Forecasting Residential Water Requirements." Journal AWWA 49, no. 3 (1957):225-235.

--------. "Peak Demands in Residential Areas." Journal AWWA 53, no. 10 (1961): 1251-1260.

WOLFF, J. B. AND J. LOOS. "Analysis of Peak Water Demands." Public Works 87 (1956):111-115.

WOLLMAN, N. The Value of Water in Alternative Uses. Albuquerque: University of New Mexico Press, 1962.

--------. Water Supply and Demand: Preliminary Estimates for 1980 and 2000. Resources for the Future Reprint No. 25. Washington, DC: RFF, 1961. (Reprinted from US Senate Select Committee on National Water Resources, Committee Print No. 32. Washington, DC: GPO, 1960).

--------. "Economic Priorities for Water Use." In Land and Water Use, edited by W. Thorne. Washington, DC: American Association for the Advancement of Science, 1963.

WOLMAN, A. "Providing Reasonable Water Service." Journal AWWA 47, no. 1 (1955): 1-8.

--------. Water, Health and Society. Selected Papers, edited by Gilbert F. White. Bloomington: Indiana University Press, 1969.

WOLMAN, A. AND H. M. BOSCH. "U.S. Water Supply Lessons Applicable to Developing Countries." Journal AWWA 55, no. 8 (1963):946-956.

WONG, S. T. Perception of Choice and Factors Affecting Industrial Water Supply in Northeastern Illinois. Research Paper No. 117. Chicago: Department of Geography, University of Chicago, 1969.

--------. "An Econometric Analysis of Urban Municipal Water Demand." 1970, 25pp. Mimeographed.

WOODS, B. M. AND E. P. de GARMO. Introduction to Engineering Economy. New York: MacMillan, 1947.

WRIGHT, JIM. The Coming Water Famine. New York: Coward-McGann Inc., 1966.

YARBOROUGH, K. A. "Analysis of Seasonal Water Consumption in Danville, Ill." Journal AWWA 48, no. 5 (1956):479-484.

STUDY AREA MATERIAL

ALLAN, L. B. "Water and Sewage Works in Metropolitan Toronto." The Municipal Utilities Magazine 92, no. 4 (1954):54ff.

BERRY, A. E. "Ontario Water Resources Act." Journal AWWA 50, no. 9 (1958):1127-1131 and Discussion, 1131-1135.

--------. "Water Resources Program in Ontario." Journal AWWA 51, no. 12 (1959): 1511-1516.

--------. "A Tribute to Norman Howard Joseph." Journal AWWA 56, no. 10 (1964): 1369-1374.

CHAPMAN, L.J. AND D.M. BROWN. The Climates of Canada for Agriculture. The Canada Land Inventory Report No. 3. Ottawa: Queen's Printer, 1966.

HAVER, C.B. AND J.R. WINTER. Future Water Supply of London: An Economic Appraisal. London, Ontario: Public Utilities Commission, January 1963.

--------. Supplementary Comments Re Economic Analysis of Future Water Supply of London. London, Ontario: Public Utilities Commission, May 1963.

HORGAN, FRANK J. "Progress Report on the Metropolitan Toronto Water Supply." Journal AWWA 56, 10 (1964):1297-1302.

JAMES F. MACLAREN (ASSOCIATES) see MACLAREN, JAMES F. (ASSOCIATES)

KAY, G.H.; A.R. TOWNSHEND; AND K.F. LETHBRIDGE. "Water Pollution Control and Regional Planning -- The Grand River Watershed." In Pollution and Our Environment. Canadian Council of Resource Ministers Conference, Montreal 1966. Background Papers, Vol. II. Ottawa: Queen's Printer, 1967.

KERR, D AND J. SPELT. The Changing Face of Toronto: A Study in Urban Geography. Memoirs II, Geographical Branch, Canada Department of Mines and Technical Surveys. Ottawa: Queen's Printer, 1965.

LONDON, ONTARIO. Engineering Board of Review. City of London Report on Water Supply. London, Ont., December 1962.

MACLAREN, JAMES F. (ASSOCIATES). Report on Water Supply and Distribution for City of London. London, Ont.: Public Utilities Commission, January 1954.

--------. Report for the P.U.C. of London on Water Supply from the Great Lakes. London, Ont.: Public Utilities Commission, August 1958.

--------. Report to the P.U.C. of London on Waterworks Development to 1985. London, Ont.: Public Utilities Commission, December 1961.

--------. Report and Technical Discussion on Functional Design of the Lake Huron Supply for the P.U.C. of the City of London. London, Ont.: Public Utilities Commission, February 1964.

McDOUGALL, H. "How do You Assess Water Rates?" Canadian Municipal Utilities: Waterworks Manual and Directory 1963:50 ff.

McGONIGAL, H.J. "Economic Aspects of Environmental Quality for Ontario." Ontario Economic Review 8, no. 2 (1970):4-9.

McIVER, IAN. Urban Water Supply Alternatives: Perception and Choice in the Grand Basin, Ontario, forthcoming.

ONTARIO, DEPARTMENT OF MUNICIPAL AFFAIRS. Manual of Assessment Values. 2nd ed. Toronto: Queen's Printer, 1954.

--------. Appraisal Notes for the Assessor. Toronto: Queen's Printer, 1964.

--------. 1967 Annual Report of Municipal Statistics. Toronto: Queen's Printer 1968. (Particularly Waterworks Section, 186 ff.)

--------. "1968 Assessment Equalization Factors of the Municipalities of Ontario." Toronto, July 1, 1968. Mimeographed.

ONTARIO, REGISTRAR-GENERAL. Vital Statistics for 1967. Toronto: Queen's Printer, 1969.

ONTARIO WATER RESOURCES COMMISSION. "Report on Water Resources Survey of County of Kent." Toronto, 1958. Mimeographed.

--------. "Report on Water Resources, Survey of County of Essex." Toronto, 1959. Mimeographed.

--------. "Report on a Water Resources Survey of the County of Peel." Toronto, 1963. Mimeographed.

--------. "Report on Water Resources Survey, County of Brant." Toronto, 1964. Mimeographed.

--------. County of York, Central Area -- Study of Water Supply and Pollution Control. Toronto, June, 1967.

--------. Brief to the Corporation of the Municipality of Metropolitan Toronto on Water Supply and Pollution Control for the County of York, Central Area. Toronto, December 1967.

209

OVERGAARD, H. O. J. "Water Problems in Southwestern Ontario." Ph.D. Thesis, Columbia University. Ann Arbor: University Microfilms, 1960.

--------. "The Emerging Problems of Water Use." Canadian Journal of Agricultural Economics 10, no. 1 (1962):10-23.

PITBLADO, J. R. "The Effects of Metering on Domestic Consumption of Water -- The City of St. Catharines." B.A. Thesis, Department of Geography, University of Toronto, 1967.

PUBLIC WORKS SURVEY. "Metro Toronto Adds Fourth Water Treatment Plant. " Public Works 101, no. 5 (1970):86-88.

STANTON PIPES (CANADA) LTD. "1967, 5th Annual Survey of Municipal Water Rates of Ontario." Water Works Digest 9, no. 1 (1967). Hamilton, Ontario.

TORONTO, BOARD OF REVIEW. Board of Review on Sewage Treatment for the City of Toronto; Majority Report by Albert E. Berry [et al.] and Minority Report by W. B. Redfern. Toronto, 1939.

TORONTO, CITY COUNCIL. Judge McDougall, et al. Toronto Waterworks Investigation Reports. Toronto: E. F. Clarke, Corporation Printer, 1887.

--------. Hering, Rudolf and Samuel M. Gray. Report on the Extension of the Water Supply and on the Disposal of the Sewage of the City of Toronto. Toronto, 1889.

--------. Hering, Rudolf and Samuel M. Gray. Report on Disposal of Sewage. Toronto, 1901.

--------. Re-Extension of Toronto Waterworks System and Waterworks Statistics. Toronto, 1914.

--------. Acres, Henry G. City of Toronto Report on Proposed Extension to the Water Works System, May 15th, 1926. H. G. Acres, William Gore, Consulting Engineers, 1926.

TORONTO, DEPARTMENT OF PUBLIC WORKS. Report on City of Toronto Water Distribution System. Toronto, 1968.

TORONTO, METRO. Gore and Storrie. Report on Water Supply and Sewage Disposal for the Metropolitan Area. Toronto, 1954.

--------. James F. MacLaren (Associates). Report and Technical Discussion on Water Supply in Metropolitan Toronto. 1957.

--------. James F. MacLaren (Associates). Supplement and Technical Discussion to (1957) Report on Water Supply in Metropolitan Toronto. 1964.

--------. James F. MacLaren (Consulting Engineers). Report and Technical Discussion to Accompany the Report on the Water Distribution System for the City of Toronto. Toronto, 1968.

TORONTO, METROPOLITAN TORONTO AND REGION TRANSPORTATION STUDY. Growth and Travel: Past and Present. First Report. Toronto: Department of Municipal Affairs, Community Planning Branch, 1966.

TORONTO, METROPOLITAN TORONTO AND REGION TRANSPORTATION STUDY. Choices for a Growing Region. Second Report. Toronto: Department of Municipal Affairs, Community Planning Branch, 1967.

TORONTO, METROPOLITAN TORONTO PLANNING BOARD. Proposed Official Plan of the M.T.P.A. Part V, Water Supply and Pollution Control. Toronto, 1964.

--------. Official Plan and Supplement of the M.T.P.A. Toronto, 1965.

TORONTO AND YORK PLANNING BOARD. Gore and Storrie. Report on Water Supply and Sewage Disposal for the City of Toronto and Related Areas. Toronto, 1949.

VAUGHAN TOWNSHIP. Hopper and Associates. Preliminary Report...Sewerage Facilities in the West Don Watershed. Township of Vaughan, October 1963.

WARNOCK, J. G. "Our Water Needs -- What Will They Be?" In Proceedings, Conference on Water Resources Management, Toronto, 1966. Toronto: The Conservation Council of Ontario, 1966.

WATT, A. K. "Our Water Resources -- What Are They?" In Proceedings, Conference on Water Resources Management, Toronto, 1966. Toronto: The Conservation Council of Ontario, 1966.

--------. "Adequacy of Ontario's Water Resources." The Canadian Mining and Metallurgical Bulletin 60, no. 664:918-922.

YOUNG, G. I. M. "A Study of the Assessment of Real Property in Ontario." Prepared for the Ontario Committee on Taxation. Unpublished M.S. 1964.